Auschwitz – Forensically Examined

Auschwitz

Forensically Examined

By Cyrus Cox

Castle Hill Publishers

P.O. Box 243, Uckfield, TN22 9AW, UK

March 2019

Cyrus Cox:
Auschwitz – Forensically Examined
First English edition, March 2019
Uckfield, UK: CASTLE HILL PUBLISHERS
P.O. Box 243, Uckfield, TN22 9AW, UK

ISBN10: 1-59148-224-0 (print edition)
ISBN13: 978-1-59148-224-6 (print edition)

Distribution world-wide by:
Castle Hill Publishers
P.O. Box 243
Uckfield, TN22 9AW, UK
https://shop.codoh.com

Typeface: Garamond.

Cover illustration: Cutaway model of Crematory II in Auschwitz-Birkenau with mass-murder scene, according to the orthodox narrative currently prevailing. Exhibit at the Auschwitz Museum. Back cover: The victims enter underground Morgue #2 ("undressing room") via a basement staircase. After undressing, they move on to Morgue #1 ("gas chamber"; front cover, bottom). Except for witness assertions, there is no evidence of the existence of the four wire-mesh columns included in this model for inserting Zyklon-B pellets (only three are visible here). After murder by gassing, the corpses are said to have been taken to the furnace room at ground level by means of an undersized freight elevator in order to be incinerated in the five triple-muffle furnaces (front cover, top).

Table of Contents

Introduction

What is widely known as "the Holocaust" is a Jewish martyrdom tale. The tale tells of six million Jews in the days of National Socialism (between the summer of 1941 and the end of 1944) who were murdered, about half of them in industrialized mass gassings. The following table shows how, according to orthodox historiography, these victims are distributed among the various so-called extermination camps.[1]

Table 1: Alleged gas-chamber victims

Auschwitz/Oświęcim	900,000*
Treblinka	900,000
Bełżec	500,000
Sobibór	210,000
Kulmhof/Chelmno	240,000
Lublin-Majdanek	unknown (thousands)*
Total:	**2,750,000 + thousands**

* The total number of Jewish casualties in these camps is higher than the figure shown due to deaths from other causes (illness, exhaustion, executions etc.)

1 Rounded means of numbers as found in October 2017 on the webpages of two of the "most-renowned" Holocaust-research facilities: the US Holocaust Memorial Museum in Washington, D.C. (U) and the Yad Vashem Museum in Jerusalem (Y):
– Auschwitz: U = "over 960,000" in total; Y = "more than 1,100,000" in total
– Belzec: U = approx. 434,500; Y = 600,000
– Sobibor: U = "at least 170,000"; Y = 250,000
– Treblinka: U = "between 870,000 and 925,000"; Y = 870,000
– Majdanek: U = "unknown"/ ""between 89,000 and 110,000," mostly victims of bad detainment conditions; Y = total Jewish number of victims "60,000" or "200,000"
– Chelmno: U = "at least 152,000," Y = 320,000

Illustration. 1: *Prof. Dr. Gerhard Buhtz demonstrates to members of the International Committee of Physicians at the Katyn mass-murder site the identification of a corpse chosen by him (State Department, Illustration 16, p. 298).*

The rest of the six million would have occurred mainly due to killings in the transitorily German-occupied areas of the Soviet Union by means of executions or by use of so-called gas vans,[2] something we will not discuss here more closely.

If one considers the enormous dimensions of the claimed mass murders, one would expect for these at least those standards of proof that apply to every murder: politically independent experts who by means of forensic methods secure traces of perpetrators, of the victims, of the offense and of the weapons used.

The Germans themselves have demonstrated precisely how this is done in time of war in the case of the Soviet mass murder of Polish officers and intellectuals at Katyn[3] and Vinnytsia:[4] With their sponsorship, they enabled experts

2 About the mass murders by the so-called *Einsatzgruppen* see Mattogno 2018, about the gas vans Alvarez 2016.

3 https://en.wikipedia.org/wiki/Katyn_massacre

4 https://en.wikipedia.org/wiki/Vinnytsia_massacre

from neutral and hostile countries to conduct the examinations freely.

The How *not* to do this, the Soviets demonstrated a little later: After the reoccupation of Katyn and Vinnytsia they conducted their own "examinations," excluding public view and participation of experts from other countries, and produced – you have three guesses – a forgery that foisted these mass murders on the Germans.

Later, after the Soviet occupation of German camps, some of these same pseudo-forensic forgers became active there as well and produced… further forgeries, for example in the case of the Auschwitz Camp, presenting "expert evidence" to the International Military Tribunal, with "forensic" methods corroborating the Soviet lie that in Auschwitz at least four million people had been murdered and burned.[5]

The Soviets spread similar lies about the Majdanek Camp as well, their "experts" having claimed that one and a half million people had been killed.[6] Today, however, the Polish Majdanek Museum propounds that around 78,000 detainees perished there – approximately 5% of the total amount that was originally claimed.[7]

These few examples should suffice to show that in general one cannot trust "forensic expert reports" by "experts" from the communist Eastern Bloc. At the end of the war, the Communists under Stalin were indeed not interested in anyone finding out the truth, otherwise they also would have had neutral experts conduct the respective examinations in public view.

In reality, the only thing the Soviets were interested in during the war was how they could produce material in order to effectively spread horror propaganda against the "Ger-

5 IMT, Vol. VII, p. 589; as to the background of the four-million propaganda number of Auschwitz, see Mattogno 2003a.

6 IMT, Vol. VII, p. 590.

7 As to the amazing story of the radical reduction of the number of victims of Majdanek, see the supplements to Graf/Mattogno 2016.

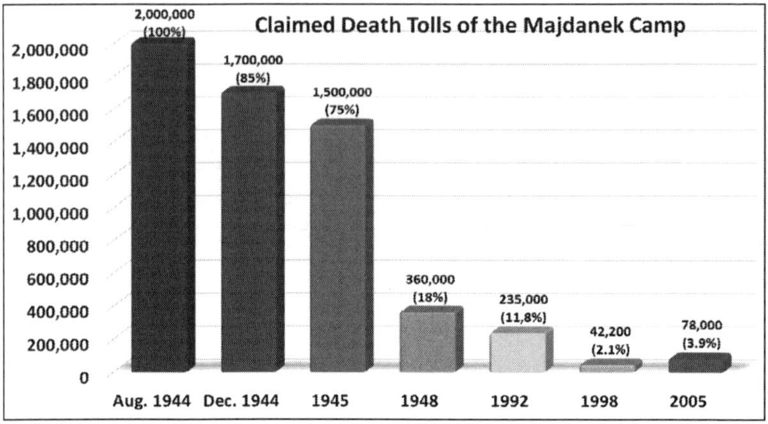

Illustration 2: *Successive reductions of the alleged number of victims for the Majdanek Concentration Camp (see Graf/Mattogno 2016).*

man-fascist Invaders" in order to wear down their morale, to spread discord amongst them, and to see to it that as many Western-leaning Germans as possible changed sides.

Furthermore, the Soviet propaganda served to divert the attention of the world from Soviet crimes, and to thwart the effective German propaganda with regard to these crimes.

After the war, the main objective was to magnify the German crimes as much as possible in order to present this as a justification for the mass expulsion and mass murder of the East and Sudeten Germans, and then later, to secure this largest land grab in human history against any German counterclaims. This was, and is to this today, plain and simple the politics of safeguarding plunder. And it was very effective and successful…

In other words: For the Holocaust story there are no sustainable proofs supported by independent forensic examinations.

At the end of the 1970s, when Dr. Robert Faurisson, at that time professor for scrutinizing texts, testimonies and documents in Lyon, France, requested such independent forensic proof, the world got to know what the Holocaust orthodoxy really thought about this inherently sound demand, because 34 French intellectuals leading in this field of study, headed by the Jewish Holocaust propagandist Pierre Vidal-Naquet, countered Faurisson's request as follows (*Le Monde*, February 21st, 1979):

> *"One cannot ask how such a mass murder was possible. It was technically possible because it happened. This is the obligatory starting point of any historical examination of this topic. This truth we simply want to bring to our memory: There is no debate about the existence of the gas chambers and there cannot be."*

Illustration 3:
Prof. Dr. Robert Faurisson

Illustration 4:
Pierre Vidal-Naquet

The Holocaust story thus had been imposed on the "civilized" world, especially on the Germans, as a commandment of faith. We were forced to believe these stories exactly because they are so unbelievable!

This brings to mind the statement by the early Christian dogmatist Tertullian, who in view of absurd dogmas wrote that it is believable exactly because of its absurd, and certain because it is impossible ("credibile est, quia ineptum est" and "certum est, quia impossibile").[8]

[8] https://en.wikipedia.org/wiki/Credo_quia_absurdum

Illustration 5: *Participants in the Auschwitz Trial of Frankfurt engage mid-December 1964 in a propaganda tour of the camp organized by the communist Auschwitz Museum. Here: A minute's silence at the "Black Wall," where executions allegedly took place. Front row, third to the left: the Stalinist expert liar Jan Sehn (Fritz Bauer Institute 2004).*

This logic of the Holocaust protagonists therefore comports with dogmatic-theocratic Holocaust theorems, but not with reason or the scientific method. It is impermissible to say a statement is true while at the same time prohibiting questions about its technical feasibility. The Holocaust orthodoxy claims that the authenticity of their Holocaust narrative is proven by the storytellers themselves. Thus, an assertion that is proven true by itself. Even if thousands of witnesses made this assertion, it still would be nothing more than an unproven claim.

After all, in the Middle Ages hundreds of thousands of alleged perpetrators, victims and witnesses claimed for instance that witches were flying on broomsticks and had wild sex orgies with the devil. Such "truth" was at the time equally "self-evident" as nowadays the claims about the Holocaust; motions to submit evidence to the courts of the Middle Ages were then fundamentally rejected, exactly as they are today during Holocaust trials, especially in Germany.

***Illustration 6:** Hans Hofmeyer, during the Second World War senior staff lawyer of the army, after that criminal judge in Frankfurt. He presided at the Auschwitz show trial at Frankfurt (Fritz Bauer Institute 2004).*

It is obligatory everywhere in the western world to credit the Jewish martyrdom story even when there is not a shred of evidence for the millions gassed. By "evidence" is meant forensic, thus laboratory-scientific findings, as the Holocaust story only leans on storytellers. Furthermore, one needs to know that even in the case of a plain murder, witness testimonies are seen as corroborative evidence only and are drawn upon if forensic examinations of the victim and of the weapon used are inconclusive or absent.

The political historiography of the Holocaust industry asserts that the Holocaust is "unique." Unique about the Holocaust, however, is only the fact that the "largest mass murder in human history" has not been forensically examined by any of the numerous courts entrusted with the punishment of so-called "National-Socialist violent criminals."

During the major Frankfurt Auschwitz Trial (1963-65) there were several on-site inspections of the former Auschwitz Camp by the court, but they merely served to manipulate the participating trial participants with false propaganda by the Polish-communist museum guides directed by Kazimierz Smoleń, at that time head of the museum. None of the trial participants thought of the idea of having exhumations of corpses or ashes conducted, or having wall samples taken from the so-called gas-chamber installations for the purpose of laboratory analysis, or even looking into the documentation of the museum archives.

With regard to the foundation on which the verdict of this trial was handed down, the judges themselves rendered a devastating assessment in their reasons for the verdict (Sagel-Grande, *et al.* 1979, p. 434):

Illustration 7: *Kazimierz Smoleń, imprisoned in the Second World War for resistance activities in Auschwitz. From 1955 until 1990 he was head of the communist Auschwitz Museum and the last protagonist of the four-million lie. Here a photo at the time of his perjury at the Auschwitz trial (Fritz Bauer Institute 2004).*

> *"The court lacked almost all possibilities of discovery available in a normal murder trial to create a true picture of the actual event at the time of the murder. It lacked the bodies of the victims, autopsy records, expert reports on the cause of death and the time of death; it lacked any trace of the murderers, murder weapons, etc. An examination of the eyewitness testimony was only possible in rare cases."*

Since then, independent investigators have repeatedly recommended such searches for clues or have even performed them themselves. To this day, however, such examinations are prevented by all possible means by officials and all judges of Germany and elsewhere. Instead of honoring such selfless efforts, these investigators were, and are, socially ostracized and in the end even prosecuted for their curious minds – for the crime of "denial."

Even today, meaningful forensic examinations could still be conducted, something that is prevented under pressure of

Illustration 8: *Dr. Ulrich Meinerzhagen, federal "German" hanging judge in Mannheim for the prosecution of peaceful dissidents of historiography.*

the Zionist lobby and their lackeys by every means imaginable.

The fact that, to this day, there is indeed no official forensic-scientific research for "the largest mass murder in human history" was confirmed by Holocaust Judge Dr. Ulrich Meinerzhagen during the trial of the German-Canadian revisionist Ernst Zündel in Mannheim in 2006. Zündel's defense attorneys had filed a motion to submit to the court the results of examinations conducted by independent researchers concerning the "murder weapon," such as the buildings in Auschwitz claimed to have been gas chambers. Dr. Meinerzhagen, however, rejected this motion on the following grounds:

> *"The motion by attorney-at-law Jürgen Rieger, filed on Dec. 12, 2006, to introduce the expert report of gas-chamber expert Leuchter and the expert report by Germar Rudolf, with which attorney-at-law Dr. Schaller agreed in full and Dr. Bock only with regard to having the Rudolf expert report read into the record, is denied, because the requested admission of evidence has no bearing on the merits of the case. Regarding such a significant and at the same time sensitive subject as the examination of the concentration of gas in the brickwork of the gas chambers, in order to get convincing results, it would be necessary to have an independent expert commission with scientists of diverse disciplines and preferably of various nationalities in order to obviate any doubt about the reliability of such an expert report."*

From a scientific point of view, Meinerzhagen is right. The possible political and societal consequences of this scientific challenge are indeed so large and so important that this matter really should be resolved by an independent expert commission with scientists of various disciplines and preferably various nationalities. The problem is, however, that the historical narrative Judge Meinerzhagen enforced per penal law is primarily based on the results of Soviet and Stalinist-Polish "expert commissions" that were the exact opposite of "independent," and whose conclusions are today exposed as lies and therefore cannot claim any reliability.

How would Mr. Meinerzhagen assemble an independent commission in light of the fact that every genuinely serious participant, following the scientific ethos of open-endedness, *must* consider it possible, both intellectually and emotionally, that the Soviet propaganda narrative turns out to be utterly wrong? Such a result is currently outlawed in 19 countries of the world!

In the eyes of Meinerzhagen and his ilk, "serious" apparently means that one doesn't question the basic accuracy of mendacious Soviet atrocity propaganda. In other words: to Meinerzhagen, "serious" means non-serious, "independent" means dependent and "truthful" means mendacious. This is called Orwellian Newspeak: War is peace, slavery is freedom, ignorance is strength…

If what Meinerzhagen implies is true, that there are no serious forensic examinations regarding Auschwitz that can be submitted to a court of law, then the State should be downright FORBIDDEN to prescribe by penal law a certain historical narrative that is not supported by solid proofs. The exact opposite is the case, however, for in the "freest state that ever existed on German soil," and in 18 other countries mainly in Europe everyone is forced by penal law to believe the Holocaust narrative dominated by Soviet black propaganda. He who does not believe, respectively he who cannot believe, must expect up to three years imprisonment in Switzerland, up five years in Germany, and up to ten years in Austria, or in special cases even up to twenty years of imprisonment, to name only a few countries that persecute historical dissidents.

A Brief History of Forensic Examinations of Auschwitz

The First Forensic Report on Auschwitz

As already mentioned, the Soviet Extraordinary Commission for the Investigation into German-Fascist Crimes regarding the "gas chambers" and crematories of Auschwitz fabricated an expert report that is not in any way based on documentary evidence or forensic examinations, but merely on fully conjured-up estimates of the gas chamber and cremation capacities. The purpose of this report was simply to "prove," come hell or high water, the made-up number of four million victims provided by Stalin. The document, issued at Nuremberg with the identifier USSR-008, is therefore not worth the paper it was written on.

The Second Forensic Report on Auschwitz

The aforementioned Soviet commission barely having finished its work, a Polish committee took up the activity of compiling a considerably more-elaborate "expert report" that would hammer even harder on the four-million-fold mass murder of Auschwitz. One of the "experts," who had already avidly lied along with the Soviet commission, the Polish Professor of Engineering Roman Dawidowski, was also significantly involved in this Polish follow-up committee. Accordingly, the results, which strictly adhered to the example of the Soviet commission, are just as false.[9]

Illustration 9: *Examining magistrate Jan Sehn, the Polish master liar about Auschwitz.*

Interesting in our context is that, on June 4, 1945, the head of this Polish investigation, Jan Sehn, sent a masonry sample to a chemical laboratory in Krakow. This sample had been taken from the room of Crematory II in Birkenau that allegedly served as a homicidal gas chamber. The sample was meant to be analyzed for cyanide residue that can form in masonry when the latter is exposed to hydrogen cyanide, the active ingredient of Zyklon B.

However, Dr. Jan Robel, who worked for this laboratory, never mentioned this masonry sample in his report of findings, which indicates that already back then it was determined that there is no significant cyanide residue to be found in this masonry.[10]

The suppression or concealment of data that go against a desired result is one of the main hallmarks of unscientific or pseudo-scientific work.

Thus, this Polish expert report on Auschwitz is not worth the paper it is written on either. It only served the purpose of supporting Stalinist show trials that provided a basis for the long-term goal of Polish land grabbing, at that time in full

---

9 The expert report is in the records of the trial against Rudolf Höss that was held in Krakow, Volume 11, pp. 1-57; for an analysis of the "evidence" submitted there, see Mattogno 2015.

10 Files of the Höss Trial, Appendix No. 12 in Volume 11a; the text of the accompanying letter as well as Dr. Robels's results are reproduced and translated to German in: Dokumentationszentrum… 1991, pp. 38-40; Bailer-Galanda *et al.* 1995, pp. 82-86.

***Illustration 10:** Walter Dejaco (left) and Fritz Ertl (right): The master builders of the crematoria of Auschwitz-Birkenau. They were acquitted on the basis of an architectural expert report.*

swing in the east of Germany by means of expelling some six million East Germans from Silesia, Pomerania, East Brandenburg, Posen, West and East Prussia.

The First Independent Forensic Report on Auschwitz

In 1972, the two architects Walter Dejaco and Fritz Ertl, who were involved in the planning and construction of the crematoria at Auschwitz-Birkenau, had to stand trial in Vienna for assisting in mass murder.[11] The Auschwitz Museum had sent the Viennese court the construction plans of these buildings. Because the judges found themselves incompetent to evaluate these plans, they tasked the Viennese architect Gerhard Dubin, a certified engineer, to examine these designs to ascertain whether the spaces denoted by the Ausch-

[11] District Court Vienna (Ref. 20 Vr 3806/64), Jan. 18 until March 10, 1972; compare Loitfellner 2002, pp. 163-168; 2006, pp. 183-197.

witz Museum as execution chambers could have been used as such or could have been restructured for such use. Dubin answered "No" to both questions in his expert report. This was one of the reasons why both defendants were ultimately acquitted by the jury. Subsequently, an unknown person removed Dubin's embarrassing (for the orthodoxy) expert report from the trial records, because today it is not to be found there.[12] This destruction of evidence is not only grossly anti-scientific, it is also a criminal act.

The Second Independent Forensic Report on Auschwitz

In the spring of 2007, the independent German-Canadian writer, political activist and Holocaust researcher Ernst Zündel, who would not be intimidated by penal law, was sentenced to a total of seven years of imprisonment by the Mannheim District Court for his historical dissent. This happened because in 1988, for a criminal trial that was staged against him in Canada for "denial," he had independently ordered a forensic expert report to be compiled about "the weapon of the largest mass murder in the history of the world," thus about the "gas chambers" of Auschwitz (and Majdanek), and subsequently he had exploited the results and their historical and political significance in a media campaign lasting some 15 years.[13]

Under the direction of already-mentioned Professor Dr. Robert Faurisson, the then US expert for execution technologies Fred Leuchter had been tasked in early 1988 with the forensic and technical examination of the buildings in

12 Private message by certified engineer Walter Lüftl, who interviewed Gerhard Dubin. See Lüftl 2004.

13 Regarding the prosecution history of Ernst Zündel see Hoffman 1995, Kulaszka 1992, Lenski 1990.

***Illustration 11:** Fred Leuchter (left), Ditlieb Felderer (middle) and Dr. Robert Faurisson (right) 1991 in Munich.*

Auschwitz and Majdanek claimed to have been homicidal gas chambers. Hereupon Fred Leuchter and his assistants visited Auschwitz and Majdanek and examined the buildings labeled as mass-murder gas chambers, as well as remains of demolished buildings. While doing so, he also took masonry samples and had them analyzed by Alpha Analytic Laboratories (Massachusetts).

Leuchter's expert conclusion was devastating for the Holocaust orthodoxy. According to Leuchter, from a functional and safety-related perspective, the facilities denoted as "gas chambers" could not be used for the claimed mass murder. And the masonry samples of the so-called mass-murder gas chambers did not contain significant residue (cyanides) of the toxic gas Zyklon B (see Leuchter/Faurisson/Rudolf 2017).

Although Zündel's judge Meinerzhagen himself wrote into the trial record, as mentioned earlier, that to date no forensic examinations of the "murder weapon" existed, he denied the motion of Zündel's defense attorneys to at least allow the Leuchter expert report to be introduced into the proceedings. An independent, respectable judge would have not convicted Zündel to seven years of imprisonment, but would have established the independent commission mentioned by Meinerzhagen and would have ordered the conduct of corresponding investigations.

The Third Independent Forensic Report on Auschwitz

One year after the compilation of the Leuchter expert report, the young German certified chemist Germar Rudolf encountered it. After a thorough reading, he evaluated it as somewhat superficial and flawed, however. For this reason, he shortly thereafter started to address the deficits of Leuchter's report by compiling a better one. In 1991, he travelled to Auschwitz in order to complete his chemical-physical investigations by means of analysis of masonry samples from the buildings in question in Auschwitz.

Rudolf was open-minded as to the results. He had made it plain to the defense attorney Hajo Herrmann, who had hired him, that he would refute Leuchter should the results of his examination demand this. But he would also not shy away from confirming the pioneer of the forensic Auschwitz research, in case the results of his research results required this. The expert Rudolf took extensive masonry samples, not only from the alleged mass-murder gas chambers and the delousing buildings, in which Zyklon B was applied exclusively to kill lice and other vermin, but also from a couple of prisoner quarters.

During his spare time, this young researcher, who at that time was a PhD student at the Max Planck Institute for Solid State Research in Stuttgart, had his samples analyzed by the renowned Fresenius Institute. The samples of the disinfestation facility, where according to official Holocaust literature exclusively the clothing and personal belongings of the detainees were fumigated, contained an extremely high concentration of residue that had formed from the toxic gas used at the time, Zyklon B. In the samples taken from the normal prisoner quarters as well in those of the alleged mass-murder gas chambers, however, there were no detectable residue concentrations. Rudolf also examined the "evaporation rate" of the toxin from the carrier material, and after thorough scrutiny arrive by and large at the same conclusions as Leuchter.[14]

Illustration 12:
Germar Rudolf, summer of 1991, at the time of the making of his expert report.

For his scientifically immaculate expert report and his subsequent research, Rudolf was brought to court twice.

In 1995, the District Court of Stuttgart, which denied a motion to have Rudolf's scientific research evaluated for its scientific soundness, sentenced him to 14 months' imprisonment because of his expert report. Before being summoned to serve his prison sentence, however, Rudolf made his way abroad and lived in the US, working as a publisher of scientific research since 1999.

There, at the end of 2005, he was unlawfully arrested by the US authorities and by means of deportation handed over to Germany's judiciary. On March 5, 2007, the unrefuted expert was sentenced by the District Court of Mannheim to

14 Over the years Rudolf has frequently updated his expert report, improving it by new research results. For the most recent edition, see Rudolf 2017a.

two and a half more years of imprisonment for his activities of publishing scientific material in the US, although these activities were fully legal there. This occurred after the court had denied the motion by the defense to evaluate the correctness or flawed nature of Rudolf's further research as summarized in his book *Lectures on the Holocaust.*[15]

Ongoing Forensic Auschwitz Research

In 1989, the Jewish Beate-Klarsfeld-Foundation, which dedicates itself to the dissemination of the Soviet-Zionist Holocaust narrative, published an enormous, 564-page book in coffee-table format, in which its author, the French amateur historian Jean-Claude Pressac tries to substantiate the claimed technique and operation of the homicidal gas chambers at Auschwitz (Pressac 1989). Pressac set out from the premise that the foundations of previous Auschwitz "research" were the mendacious Soviet and Polish "expert reports" of the immediate post-war period that were completely unfounded and therefore useless as guidelines.

Pressac promised to solve the problem in his book in a technical manner. With this book as well as with the one he published four years later (Pressac 1993/1994), he missed that goal abjectly, however, as his books do neither contain even one reference to expert literature about cremations nor to the usage of toxic gases. He did not perform his own calculations or experiments either.[16] He merely took some wartime documents out of context and built his tales of horror around them, although these tales are supported only by "witnesses," most of whom testified before the Soviet

---

[15] About Rudolf's story of prosecution, see Rudolf 2016a&c. For the most-recent edition of his *Lectures*, see Rudolf 2017b.

[16] For a critique of Pressac's total failure, see Mattogno 2015 as well as Rudolf 2016b.

and/or the Polish commissions with the clear purpose of shoring up the four-million lie.

Illustration 13:
Jean-Claude Pressac in 1993.

Two Italian researchers ultimately filled this persistent gap: the historian Carlo Mattogno and the engineer Dr. Franco Deana. Immediately after the publication of Pressac's first book that allowed third parties to form their own opinion on the basis of the reproduction of numerous documents, these two researchers set about to put the question with regard to the technology, operation and capacity of the crematoria and gas chambers of Auschwitz on a truly solid foundation, substantiated by documents and technical and exact scientific evidence. The first glimpse the public got into their progressing research came from an article in an anthology concerning the Holocaust published in 1994 (Mattogno/Deana 1994).

Shortly afterwards, Mattogno and colleagues gained access to the archival material of the former Auschwitz camp authorities stored in Moscow. Because of that, and also due to other new archival discoveries, knowledge of what transpired at Auschwitz increased enormously, and their studies of the crematoria expanded substantially.

After 25 years of research, their opus magnum was finally published in English in 2015 (Mattogno/Deana 2015). This monumental work of 1,200 pages in three volumes concerning the cremation furnaces of Auschwitz contains references to a large number of professional articles about the history and technology of cremations, abundant documentation concerning the crematoria of Auschwitz and other camps, and elaborate thermo-technical calculations, as well as references to a panoply of experiments.

Illustration 14:
Carlo Mattogno

The conclusions of both experts are clear and distinct: The crematoria of Auschwitz were downsized versions of civilian crematoria, barely capable of incinerating those victims who had fallen victim mostly to epidemics raging in Auschwitz (mainly typhus). In no way would they have been able additionally to incinerate the alleged victims of mass extermination by poison gas.

A further aspect of the Auschwitz narrative that has inspired forensic research are allegations by many witnesses about burning corpses on pyres outdoors, which were said to have been erected in deep pits in the vicinity of the Birkenau Camp.

In this context, two papers are worth mentioning with regard to the question of whether incinerations in deep pits would have been possible at all, given that the Birkenau Camp was situated precisely at the confluence of two rivers, the Sola and the Vistula. One of the papers was authored in 1998 by the German architect Willy Wallwey (under the pseudonyms Michael Gärtner and Werner Rademacher), the other four years later by the tireless Italian historian Carlo Mattogno (Mattogno 2002).

Both came to the same conclusion: The groundwater in Birkenau was on average so close to the surface that parts of the area were marshy. Pits with a depth of more than a meter (3.3 ft) would have swiftly filled with groundwater. In the face of this, neither deep mass graves nor deep incineration

Illustration 15: *Auschwitz-Birkenau in the rain: inundated.*

pits would have been possible. The present work does not deal with this theme any further, however.[17]

What certainly will be discussed here, however, is the general question of whether cremations on pyres outdoors took place in Auschwitz, and if so, to what extent. A first paper about this with a forensic approach was published in 2004 by Heinrich Köchel (updated in Köchel 2016). Köchel's article was shortly thereafter expanded upon by a comprehensive study of this question in English (Mattogno 2005, updated in Mattogno 2016b).

The results of this study were again unequivocal: The allegations of witnesses about incinerations outdoors of hundreds of thousands of corpses in 1944 were conclusively rebutted by aerial photos, taken over Auschwitz by reconnaissance aircraft starting in May 1944.

Unlike the papers mentioned in the previous chapters, none of the studies mentioned here was submitted to any court as evidence. They did not come into existence with the

[17] Both articles are reprinted in Part Three of Mattogno 2016b, pp. 97-127.

aim of deciding a court case but out of pure scientific inquisitiveness.

* * *

Are the expert reports and scientific studies by Gerhard Dubin, Fred Leuchter, Germar Rudolf, Carlo Mattogno and Franco Deana, Willy Wallwey and Heinrich Köchel sound or false? Were Rudolf and Ernst Zündel as well as so many other truth-seekers wrongly prosecuted and imprisoned for years? In the following chapters, we will pursue this question.

The Chemistry of Auschwitz

The Historical Framework

The reason for the existence of the enormous camp at Auschwitz was primarily the lack of manpower in the Third Reich during the war – more and more men were at the fronts – and secondly the geographically favorable location of the place.

Auschwitz is near the confluence of the Rivers Sola and Vistula, not far from the coal mines in Upper Silesia; it's a railway junction between the East and the West, the North and the South; and until the end of 1943, it was far beyond the reach of the bomber fleets of the Western Allies. This was the reason why the planners of the Third Reich decided to develop coal-based chemical industry there. The heart of this large-scale project was the BUNA Works, a coal-refining plant intended to synthesize raw material from coal for the chemical industry as well as ultimately gasoline, kerosene and artificial rubber. These processes required large quantities of water, which was planned to be taken from the Vistula.

In order to remedy the lack of manpower in the industrial area, it was attempted initially to deploy Russian prisoners of war, for whom an enormous camp in Birkenau was designed. Due to logistical bottlenecks at the eastern front, however, it turned out to be impossible to transfer larger numbers of these POWs westward.

Hence in early 1942, the Jews in Europe became the focal point. It was planned that, from March 1942 on, they would

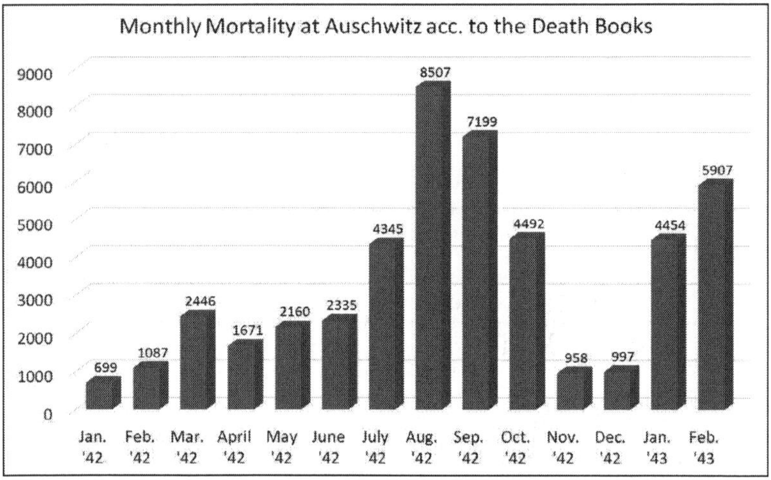

Illustration 16: *Deaths by month according to the Death Books of Auschwitz (State Museum 1995).*

fill in for what was originally planned for the Russian POWs to do.

Until fall of 1942, the Birkenau Camp, which was being hastily expanded by prisoners, lacked adequate washing, toilet and delousing facilities. Consequently, the hygienic situation in the camp was catastrophic. Lice, fleas and bugs spawned at an explosive rate. In March 1942, typhus appeared in the camp due to infected lice. Since that time the death rate in the camp rose continually, and surged in July and August to such catastrophic heights that the entire Auschwitz camp complex was locked down in quarantine. Towards the end of the year, the epidemic appeared to have been brought under control by means of massive deployment of Zyklon B against the lice, but in early 1943, a new outbreak suddenly emerged. See Illustration 16.

From July until October 1942, and again from January 1943, hundreds of detainees died daily in Auschwitz due to the epidemic. The new crematoria of Birkenau were only built starting in summer 1942, and were not operative until March 1943. The small, old crematory in the Auschwitz

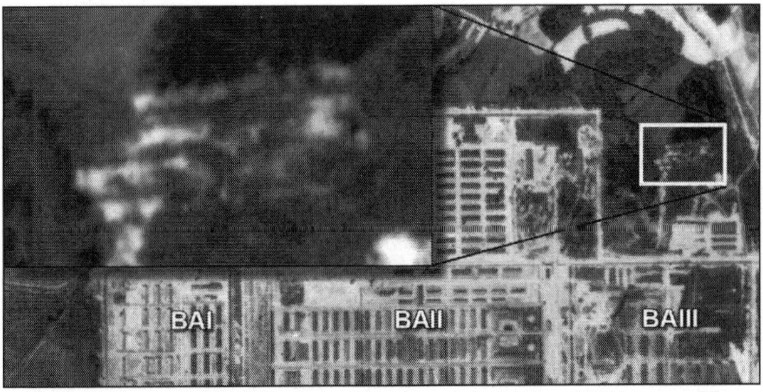

***Illustration 17:** Detail of an Allied aerial photo of Birkenau May 31, 1944. The four elongated rectangles in the image top right, enhanced in the cutout to the left, were probably mass graves of victims of the catastrophic typhus epidemic of 1942/1943 with a maximum capacity of approximately 20,000 corpses. BAI to BAIII: The three sectors of the Birkenau Camp (see Rudolf 2018, p. 119).*

Main Camp could not cope with the load on its own. It had become necessary to temporarily dispose of the corpses elsewhere.

Initially, starting in summer 1942, it was decided to bury the corpses that could not be incinerated in shallow mass graves near the Birkenau Camp (see Illustration 17). But then it became clear that the corpses were submersed in the groundwater, with the threat of polluting the water supply. As the whole region drew its drinking water from this groundwater, the corpses had to be exhumed. This absolutely horrific work was done by detainees starting in late summer 1942. The bodies of all the typhus victims who died from late summer on and who could not be cremated were immediately burned on pyres. This only ended when the completion of the crematoria of Birkenau enabled the handling of mass mortalities.

The massive deployment of Zyklon B in the battle against the lice with a simultaneous mass mortality of detainees due to typhus, the mountains of malodorous corpses caused by the exhumation of partly decomposed victims of typhus with the accompanying stench of decomposition as well as the pyres that must have appeared infernal, are the true core of the holocaust stories of Auschwitz.

Fact is, however, that in this period the camp authorities weren't logistically able to dispose of any victims of mass extermination atop the often hundreds of corpses daily that were already impossible to dispose of.[18] A series of documents exists about this which proves that in 1943 and 1944 all morgues of the crematoria of Auschwitz-Birkenau were available day and night to be used as storage places for corpses of detainees who had died of all causes in the entire camp (Mattogno 2004b). Therefore, at no point in time could these facilities have served as locations of mass murder as the orthodoxy claims.

Zyklon B and the Consequences

Since the 1920s, Zyklon B was a trademark of the *Deutsche Gesellschaft für Schädlingsbekämpfung* (DEGESCH, German Company for Pest Control).[19] Besides its official purpose as a pest-control fumigant, this agent was allegedly used in Auschwitz for the mass murder of Jews since early 1942.

The commercial product Zyklon B, the way it was used in the German area of control during the Second World War, was essentially liquid hydrogen cyanide absorbed on gypsum pellets.[20] For storage and transport, these pellets were kept in tin cans. Until the introduction of DDT towards the end of the war, it was the most-effective of all known pest-control agents.

---

18 For the details of the camp history, see Rudolf 2017a, pp. 53-67.

19 For the history of Zyklon B, see Kalthoff/Werner 1998.

20 We limit ourselves here to the product named Erco that was used in the camps. For details, see Lambrecht 1997, Mazal.

Illustration 18: *In August 1976 this Evangelical church in D-96484 Meeder-Wiesenfeld (top) was gassed with Zyklon B. Subsequently the plaster discolored everywhere blotchy blue (Illustration source: Wikipedia commons).*

Since mid-1920s, exterminators increasingly used it for pest control. The pellets were scattered into the sealed space to be treated. In the absence of mechanical ventilation, it took about a day until the fumigated space was fully ventilated

Without doubt Zyklon B was used in large quantities in the concentration camps of the Third Reich in order to improve the hygienic conditions there, thus to protect the life and health of the detainees.

In summer 1942, the Auschwitz camp authorities built two large hygienic buildings that had a wing designated as Zyklon-B disinfestation chambers and were then used as such. That's how the walls of these chambers were exposed, over an extended period, to the active ingredient of Zyklon B – gaseous hydrogen cyanide. Due to this exposure, the walls of these chambers were discolored from the inside, penetrating to the outside with blotches of blue, as one can see from the many color images in Rudolf's book on this subject (Rudolf 2017a).

From building-damage incidents caused by the inept use of Zyklon B in civilian buildings, it is known that this blotchy-blue discoloration is indeed the result of the interaction of hydrogen cyanide with components of masonry.[21] Ultimately, it's not the blue color that is decisive, but the chemical analysis with regard to residue of hydrogen cyanide, called cyanides.

Illustration 19: *In 1972 the Catholic church in Untergriesbach was fumigated with Zyklon B. Subsequently, the plaster was discolored by blotches of blue. (Illustration source: Konrad Lackerbeck; Wikipedia commons).*

Thus, if significant quantities of cyanide are found in masonry samples of those chambers at Auschwitz that are said to have been used as homicidal gas chambers, then this would be a distinct proof for the deployment of the toxic gas in these localities.

If, however, no significant quantities are found, then there are two possibilities:

1. Mass gassings happened there anyway, but due to unknown reasons no traceable residue has been able to form there; or
2. no mass gassings occurred there.

Analyses and Subterfuges

So, first of all it needs to be determined what is found in these walls. To date there are analytical results known from

[21] See the case of a church fumigated in 1972 with Zyklon B: www.pfarrei-untergriesbach.de/pfarrbrief11.htm (accessed on July 17, 2018), as well as a second case of fumigation of a church in 1976: Zimmermann 1981, pp. 120f.

Table 2: Ranges of results of analyses of various samples, in mg CN^-/kg

Author:	Markiewicz	Leuchter	Rudolf	Ball
Detection of:	Cyanide without Iron Blue	——— Total cyanide ———		
Disinfestation	0 – 0.8	1,025	1,000 – 13,000	2,780 – 3,170
Homicidal gas	0 – 0.6	0 – 8	0 – 7	0 – 1.2

four researchers, or research groups. The following table shows the ranges of their analytic results of the masonry samples of two categories of chambers: disinfestation chambers (first row) and alleged homicidal gas chambers (second row).[22]

The results of the individual analyses are listed in Table 5 on p. 55.

Two remarks are necessary to understand this data:

1. Carbonates simulate the presence of cyanide in a sample. A main component of concrete, cement, mortar and plaster is chalk, chemically: calcium carbonate. Series of tests have shown that quantities present in such masonry samples can simulate analytic values of up to 10 mg cyanide per kg sample material.[23] Analytic values up to this rate with regard to samples that consist of such materials are therefore to be considered insignificant.
2. Markiewicz and colleagues applied an analytic method that is not capable of detecting long-term-stable cyanide compounds of the Iron Blue type – the blue pigment in question in the walls. Their method can only detect unstable cyanide compounds that, decades later, may be expected to have largely evaporated. Thus, the analytic values of such methods can be ascribed chiefly to carbonates.

---

22 For the various analyses conducted and their evaluation, see Rudolf 2017a, Chapter 8, pp. 297-359.

23 For this, see Meeussen/Temminghof *et al.* 1989 or Rudolf 2017a, pp. 299-301.

Illustration 20: *Blotchy blue discoloration of the inner walls – from the floor to the ceiling – of the Zyklon B disinfestation wing of Hygiene Building 5a in Birkenau, Construction Section I (© Carlo Mattogno 1992).*

The analyses of Markiewicz and colleagues published in 1994 were compiled at the same forensic institute in Krakow where in 1945 examining magistrate Jan Sehn and Prof. Dr. Dawidowski compiled their mendacious propaganda expert report about Auschwitz. Today this institute carries the name of this monumental fraud: The Jan Sehn Institute for Forensic Research.

Illustration 21: *Dr. Jan Markiewicz, Polish professor for sleight of hand.*

Markiewicz was specifically tasked by the Auschwitz Museum to rebut the findings of Fred Leuchter and Germar Rudolf. In order to achieve this desired result, right from the start

Illustration 22: *Stained blue discoloration of the outer walls of the Zyklon-B-disinfestation wing of the hygiene Building 5b in Birkenau, Construction Section I (© Carlo Mattogno 1992).*

those very cyanide compounds were excluded from the analysis that alone could be found 50 years after the exposure. In this way in both sample types – those from the disinfestation chambers and those from alleged homicidal gas chambers – practically nothing was found. From that the crafty Poles concluded that both locations must have had a similar history. Because a large amount of Zyklon B had been used in the disinfestation chambers, which is undisputed, they made the conclusion that in the premises tagged as homicidal gas chambers, a lot of Zyklon B must have been used as well.

In other words: According to the Poles, the absence of unstable cyanide compounds in the masonry of the alleged homicidal gas chambers proves the massive use there of Zyklon B. That's a new Polish logic: The non-existence of something proves the existence of it!

One rubs one's eyes in wonderment: Why have the Poles stated such nonsense?

Illustration 23: *John C. Ball, the third sample taker and author of the groundbreaking book* Air Photo Evidence *about air photos of Auschwitz and other crime scenes of the Holocaust (now Rudolf 2018).*

They came up with many dubious reasons, for instance that they could not understand how blue pigments could form in walls after exposure to hydrogen-cyanide gas. For that, one should tell them that, if they don't understand their trade, they should quit their job! Or they thought that blue pigments could stem from blue wall paint which they did not want to detect. But in the chambers in question, there is not a trace of blue paint – for pigment is only a small part of paint. The bulk of paint consists of a binding agent and all kinds of chemical ingredients. Furthermore, no wall paint exists that can be found in the mortar situated in the middle(!) of the wall 40 years later, because the gas has penetrated the wall and formed this pigment everywhere: inside, outside, in the middle. How could that be achieved using wall paint? And in any case, why would the SS have painted the delousing chambers blue, particularly in a blotchy random pattern?

The cockeyed blabber about not understanding or about blue wall paint is egregious flimflam at best; in reality, however, it is a lie to confuse the audience. Already the aforementioned cases of damage to churches prove this, as do the walls of the disinfestation chambers from the times of the Third Reich, saturated through and through with the blue pigment. And the Poles knew that, because in their article they quoted a book in which all of this was explained and proved by expert literature (Gauss 1993, pp. 163-170; 290-294).

Let us not be naïve; the facts are obvious for anyone willing to see: Markiewicz and colleagues upheld the tradition of the Polish "experts" of their Institute in order to disseminate the baldest lies about Auschwitz. After all, the Auschwitz Museum requested that report from them, and "Auschwitz denial" is punishable by up to three years' imprisonment in Poland! In such a case, they don't think twice: the analysis results are manipulated until they fit!

The other three sample takers who had their samples analyzed for the *total* concentration of cyanide, thus unstable *and* stable compounds, are basically in accordance with each other: While there is an enormous presence of cyanide residue in the masonry of the disinfestation chambers, in the alleged homicidal gas chambers there is no significant presence at all.

Obsessive Explanation Attempts

As a matter of survival, the proponents of the orthodox Holocaust story *have to* explain away these clear analytic results from the alleged homicidal gas chambers. After all, in their eyes the gassing of humans is rock-solidly proven by the testimonies of "witnesses." To the orthodoxy, forensic research results that don't confirm these testimonies are nothing other than a thorn in the flesh, along the lines of: I've got my opinion, don't bother me with facts!

Dr. James Roth, who had supervised the analysis of Fred Leuchter's samples at the Alpha Analytic Laboratories without knowing what they were about, was later cornered by a journalist for aiding "denial." To "exonerate" himself, Dr. Roth thereupon made up the fairy tale "that cyanide will react on the surface of […] plaster, penetrating the material not more than 10 microns, or 0.01 mm [0.0039 inch], or one

***Illustration 24:** James Roth during an interview in which he lied through his teeth (Morris, 1:03:41).*

tenth the thickness of a human hair" (van Pelt 2002, p. 390). That was a lie. To hydrogen-cyanide gas, plaster and mortar are as permeable as a sponge is to water. Experimental data of gassing experiments from 1929 (Schwarz/Deckert 1929) as well as general data from building material irrefutably prove this (see German Norm DIN 4108, Part 4). Roth knows this, because when he testified under oath at the 1988 Zündel trial, he truthfully said: (Kulaszka 1992, p. 363):

> "*In porous materials such as brick or mortar,* [hydrogen cyanide] *could go fairly deep as long as the surface stayed open* […]."

More streetwise with his obsessive explanation attempts was the French biochemist and Auschwitz survivor Dr. Georges Wellers. His opinion was that humans are considerably more sensitive to hydrogen cyanide than insects, which was allegedly the reason why homicidal gassings were conducted with smaller amounts of hydrogen cyanide in shorter times. During that short time, the victims would have inhaled almost all of the hydrogen cyanide, so that presumably there was nothing left to react with the masonry (Wellers 1991).

It is indeed true that warm-blooded creatures such as humans are considerably more sensitive than insects, but Wellers overlooked several things:

1. Experience with executions by means of instantly released hydrogen cyanide in the execution gas chambers of the US shows that in these cases of applying hydrogen-cyanide concentrations similar to those used against in-

sects, it took on average around 9 minutes before the victims were dead, and in extreme cases up to 18 minutes (Christianson 2010, pp. 220, 229).

Illustration 25:
Dr. Georges Wellers during his perjurious witness testimony at the Jerusalem Eichmann trial.

2. In Auschwitz, Zyklon B was used, which only slowly discharged its toxin over a period of one to two hours (Irmscher 1942). Moreover, in *none* of the homicidal gas chambers alleged for Auschwitz did devices exist to aid evaporating the hydrogen cyanide (such as a warm-air blower) and thereafter spreading the toxic gas in the large gas chambers (for instance an air-circulation blower). Such devices were part of the standard equipment of the professional disinfestation chambers mass-produced in that period. In Auschwitz, however, the toxic pellets are said to have been simply dumped into the chambers or lowered into them in baskets.
3. The only witnesses who, with regard to the alleged gassings in Auschwitz, could have made reliable statements as to the deployed quantities of the poison and the duration of the executions were the SS physicians who controlled the issuance of Zyklon B and who allegedly had carried out and supervised the mass murders. In this regard, statements of only four of the about twenty known physicians posted at Auschwitz are known, and all of them mentioned execution durations of two to five minutes until the death of *all* victims, but made no statements as to the amounts of Zyklon B used (three of them are shown in Illustrations 26f.; see Rudolf 2017a, pp. 254-257).

4. These claimed execution durations, according to which *all* victims would have been dead – including the most-uncooperative ones, who simply again and again held their breath, the most-fit ones who weren't easily taken down by anything, and those who stood the farthest away from the Zyklon B – are much shorter than those measured at executions in the US where the victim was *immediately* exposed to the *full* gas concentration. Accordingly, in the homicidal gas chambers of Auschwitz there must have been toxic-gas concentrations on average that were correspondingly higher. To achieve this with Zyklon B without auxiliary devices (warm-air blowers and circulation-air blowers), is practically impossible for the extreme values (2 or 3 minutes), and for the longest-claimed duration (5 minutes) would have required the application of a gigantic quantity of Zyklon B (see Rudolf 2017a, pp. 259-269).

Illustration 26:
Dr. Hans Münch, physician in Auschwitz (July 1997)

Illustration 27:
Dr. Johann Paul Kremer, physician in Auschwitz (June 4, 1964).

Illustration 28:
Dr. Horst Fischer, physician in Auschwitz, during his trial in communist East Germany.

5. Zyklon B was reportedly simply dumped into the chambers. The murder is said to have been completed after 5 minutes at the most, but the Zyklon B at this time would have only discharged a small fraction of its hydrogen cyanide. The concentration of toxic gas in the chambers would thus have steadily increased for one to two hours, depending on the temperature. Ventilation of the chamber before the complete evaporation of the hydrogen cyanide would have been of no avail.
6. In this scenario, the victims, before dying, could have inhaled only an insignificant part of the hydrogen-cyanide gas that was in the chamber.

Hence, it is clear that the claimed mass murder, the way it is said to have happened, could not have been committed with small quantities of poison gas. Exactly the opposite is true. It's also clear that the duration of the claimed procedure would have been significantly shorter on occasion than the duration of disinfestation gassings which, in the absence of auxiliary devices (a warm-air blower and air-circulation blower) could last up to 24 hours before ventilation could be started.

With this, factors come into play that are intentionally disregarded by orthodox scholars. The claimed homicidal underground gas chambers of Crematories II and III of Auschwitz-Birkenau that were officially marked "Morgue #1" – the only locations remaining from where samples can be taken in a useful manner[24] – had a significantly higher tendency of forming long-term-stable cyanide residue than

24 The walls of Crematories IV and V were completely destroyed by the demolition of the buildings at the end of the war and the fragments were "taken care of" by the museum. The walls of the so-called Bunker 2 do not exist anymore, and the putative Bunker 1 never existed. The walls of Crematory I in the main camp were then already decades old (= hardly reactive) and were allegedly only exposed a few times to toxic gas.

Table 3: Difference between the absorption of hydrogen cyanide (HCN) in moist and in dry material (absorption of hydrogen cyanide at exposure of 2 Vol.-% HCN during 24 hours; Schwarz/Deckert 1929, p. 201).

MATERIAL	HCN [mg m^{-2}]
Sand-lime brick, naturally moist	22,740.0
Sand-lime brick, dried during ca. ½ a year at 68°F	2,941.0

the disinfestation buildings, which still exist to this day in the same camp (see the summary in Rudolf 2017a, pp. 353-357):

1. For pigment formation, a high degree of humidity in the masonry is beneficial, see Table 3. This is essentially a function of temperature. Cold walls are moist, while warm walls are dry. The underground morgues in question were unheated and therefore had constantly cool, moist walls. The disinfestation chambers were aboveground and were constantly heated during their use. Their walls were therefore dry. The difference can amount to up to a factor of eight.

2. For pigment formation, a low acidity of the masonry is beneficial. The underground morgues had mortar and plaster rich in cement. This keeps the acidity of the masonry low (alkaline) for years. The disinfestation chambers had plaster that consisted of lime which is free of, or low in, cement content and had a mortar between the bricks that was equally low in cement. Within a few days after applying the mortar or plaster, the acidity of these materials increases rapidly and soon reaches neutral values (setting of the lime). The difference can be a factor of ten or higher, see Table 4.
3. For pigment formation, a high inner surface of the masonry is beneficial. The cement mortar of the underground morgues has a significantly rougher and larger microscopic surface than the plaster of the disinfestation chambers (see Illustration 29). Rough, large surfaces ac-

Table 4: Cyanide concentration in gassed samples in mg HCN per m² exposed surface 90 hours after the termination of the gassing (Schwarz/Deckert 1929, p. 203)

Material	mg
Cement mortar, dry, hardened (moderate acidity)	36.2
Concrete block, mildly fresh (low acidity)	1,926.0
Factor difference	53

The samples were exposed during 24 hours to a nominal concentration of 22.5 g/m³ hydrogen cyanide; the actual concentration steadily decreased, however, due to the absorption by the samples. After that, the concentration in the hardened cement sample steadily decreased, but did not in the fresh concrete sample.

celerate every chemical reaction. The difference can be up to a factor ten or higher.

In order to rate the combined effects of these factors, one cannot unconditionally multiply them by each other. The fact is, however, that the chemical and physical tendency of the underground morgues to accumulate stable cyanide residue would have been so significantly higher that this would have more than sufficed to compensate for the shorter gassing duration assumed for homicidal gassings compared to fumigations.

In other words: In the masonry samples of the underground morgue, we should find approximately similar cyanide residues as in the disinfestation chambers, if not even more, *provided* that the stories told by the witnesses are true.

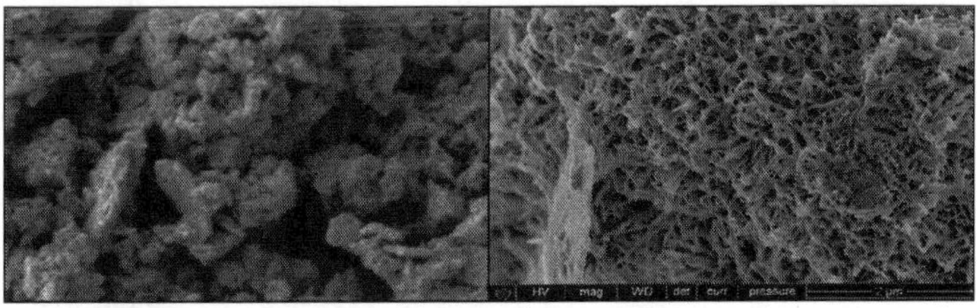

Illustration 29: *Images by a scanning electron microscope of calcium mortar (left) and cement mortar (right) with comparable magnification. The inner surface of the cement mortar is much rougher and larger than the one of the lime mortar.*

Absurd Auxiliary Theses

In order to solve the whole problem, it would have become the Soviet and Polish "experts" of the investigation commission to apply the technical standard of advanced German disinfestation facilities to the alleged chemical slaughterhouses – including warm-air blowers and air-circulation blowers, see Illustration 30 – and to instruct the witnesses to that effect.

After all, the Auschwitz camp authorities knew about this standard, because a description of this technology was sent to them in 1941 (it's in the Auschwitz archives to this day), and the latest Zyklon-B disinfestation facilities that were based on it had been planned for the reception building at the Auschwitz Main Camp.

In the end, however, an even more-modern facility was built instead: The first microwave oven in the world – for the killing of harmful insects in the detainees' clothing and bed linen (see Nowak 1998, Lamker 1998, Nowak/Radema-

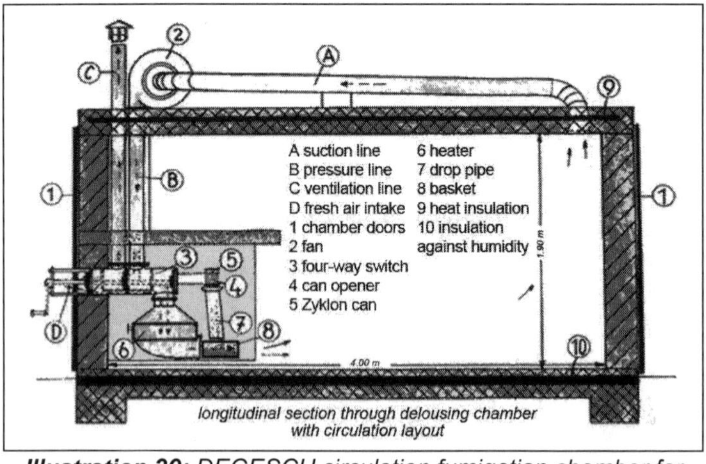

Illustration 30: *DEGESCH circulation fumigation chamber for the professional application of Zyklon B. This standard had to be expected for hypothetical execution chambers.*

cher, pp. 312-322). Therefore, the Auschwitz detainees never knew about the technology of advanced German Zyklon-B disinfestation facilities, and that is exactly what their testimonies reflect.

The Soviet and Polish "experts" weren't the cleverest either, because instead of informing themselves about the technology, they only required the witnesses to somehow slant their stories with absurd statements supporting the four-million lie.

Since neither the members of the investigation commission nor the witnesses had the slightest clue about German Zyklon-B delousing technology, the whole propaganda project had to ultimately derail. That nearly all witnesses claimed the Zyklon-B pellets had been poured through some hole or another in the walls or ceilings – although it couldn't have worked that way at all – clearly points to someone having coordinated the testimonies in a deliberate and amateurish way.

In order to somehow extract themselves from this disaster, ever since Pressac the orthodoxy has focused on claiming that, in the underground Morgue #1 alias "gas chambers" of Crematoria II and III of Birkenau, some hollow wire-mesh columns had been installed, in which Zyklon B had been lowered in some manner and had been hoisted back up again after completing the murder.[25]

Too bad that for these claimed columns neither documentary nor physical traces exist; that absolutely indispensable anchoring points for these columns in the ceilings and floors are nowhere to be found; that the witnesses who claim the existence of the columns contradict each other with regard to the operation mode; and that the most important witness for these columns, who claims to have been involved when they were being built, described them under oath –

---

25 See Pressac 1989, p. 487; van Pelt 2002, pp. 194, 208; and McCarthy/van Alstine.

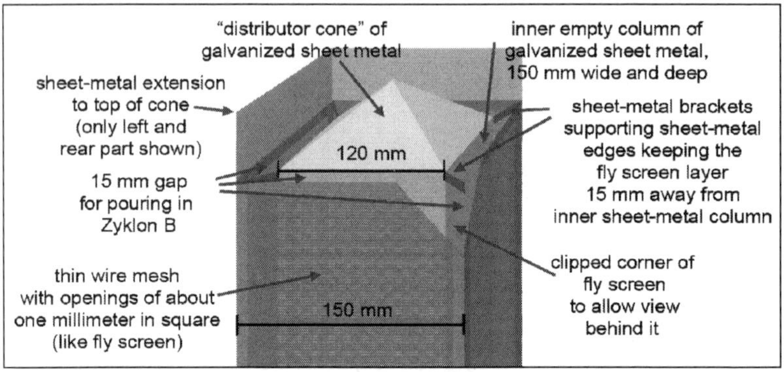

Illustration 31: *Inner part of the Kula column, according to his statement under oath at the Höss trial (Drawing by the author).*

during the trial of the former Auschwitz camp commander Rudolf Höss – in such a way which would have made it impossible for them to function:

Witness Michał Kula opined that Zyklon B had been poured into an inner, removable column that consisted of two layers: an inner layer made of sheet metal, sealed on top by a conical cap, and at a distance of 15 mm (0.59 inch) an outer layer made of screen with a mesh size of 1 mm (0.039 inch). Into the gap between the two layers, Zyklon B was supposedly poured. See Illustration 31.

The Zyklon-B gypsum pellets were some 0.5 to 1 cm in size (0.2 to 0.39 inch), see Illustration 32. During the pouring-in of these pellets into the 0.59-inch-narrow gap, the one side of which consists of screen and the other of smooth sheet metal, it necessarily must have come to a clogging of the pellets at the top of the gap. According to expert literature, at least a three- to sevenfold clearance of the maximum particle size is needed, so here at least 1.18 inch (Schulze 2006, p. 302), for a steady flow of the bulk solids through a narrow *smooth* opening. Here, we are dealing with a very-rough mesh surface on one side, however, the reason why the size of the gap had to be at least 2.36 inch for an unimpeded flow to occur.

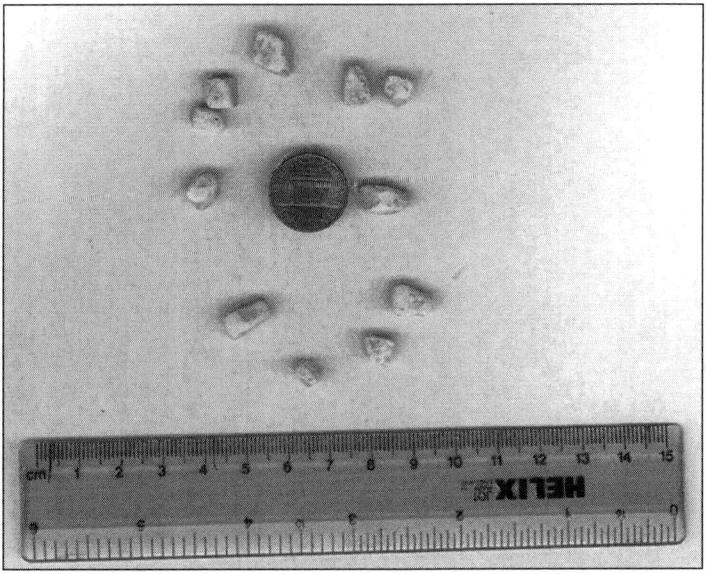

Illustration 32: *Size of the gypsum lumps in Zyklon B (see Mazal)*

So, the gap would have been immediately clogged, and it would have been impossible to introduce more Zyklon-B pellets without any further ado. One would have had to constantly knock on the device in order to get the pellets to fall. That would have warped the screen, which would have caused the gap to become even narrower and more irregular, causing the clogging tendency to increase even further. Indeed, one would have had to fill up this column while removed from the other columns, as only this would have enabled the knocking against it. This would have taken many minutes while the toxic gas would have freely enveloped the executioners trying to fill the device.

A complicating factor in this scenario is that the Zyklon B would have been introduced into a space that was saturated with water vapor. The evaporation of the hydrogen cyanide would have cooled the pellets down. Consequently, atmospheric moisture would have condensed on the gypsum pellets. That would have made the pellets sticky, causing them to be practically impossible to get out of this 0.59-inch gap after the deed was done. The column would have been dented and bent, rendering it unusable after its first use.

Hence, already from a mechanical point of view, Kula's columns wouldn't have worked. But they wouldn't have worked toxicologically either, because if the pellets are kept away from the people instead of spreading them out among them, hence if they are confined behind a closely meshed screen, the evaporation process of the hydrogen cyanide proceeds very slowly. In addition, the just-mentioned atmospheric moisture condensing on the gypsum pellets would have slowed down the evaporation process of the hydrogen cyanide right from the start so strongly – hydrogen cyanide is readily soluble in water – that the claimed execution times (2 to 5 minutes) can be fully and categorically dismissed as impossible when using Kula's columns.

Game over.

Before we turn to the next topic, a final word about the method applied in this case by orthodox Holocaust dogmatists. In logic there's the principle of

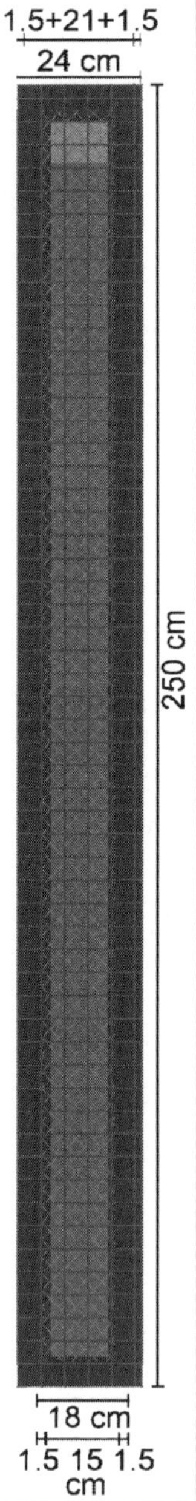

Illustration 33: *Drawing of the Zyklon-B insertion column as described by Michał Kula at the Höss trial: three nested screen columns with a progressively narrowing mesh. An enlarged view of the inner column can be seen in Illustration 31.*

Ockham's Razor, also simply called the principle of parsimony.[26] It says that, of multiple possible explaining theories for one and the same situation, the simplest theory should be preferred, and that a theory should have as few variables and auxiliary hypotheses as possible. If a theory is only able to hold its ground against another one by being shored up with auxiliary hypotheses, this indicates that it is probably false.

In our case, we have a simple situation: in the Zyklon-B disinfestation chambers of the Third Reich plenty of cyanide residue can be found; in the alleged Zyklon-B homicidal gas chambers of Auschwitz however, we find none.

The simplest explanation for this is that there were no gassings with Zyklon B in the alleged homicidal gas chambers, plain and simple.

But because in the eyes of the dogmatic orthodoxy something cannot be true that is not allowed to be true, they bolster the thesis of the existence of Zyklon-B homicidal gas chambers by all sorts of auxiliary hypotheses. One after the other was exposed as being incorrect, though. For example, the thesis about the wall paint; about the impossibility of the formation of long-term-stable residual compounds; about the impossibility of hydrogen cyanide deeply penetrating masonry; and about the short persistence of hydrogen-cyanide concentrations in the alleged homicidal gas chambers. This auxiliary thesis is the most persistent one.

The conditions claimed by witnesses of these executions require, however, that enormous quantities of poison gas had to have been used. As this would then have aggravated the ventilation problem, a second auxiliary thesis is immediately brought forward to bolster the first one: Well, then devices existed that allowed for the removal of Zyklon B from the chamber so that they could be ventilated quickly. This auxiliary thesis also goes to pieces when scrutinized, however, as I just demonstrated.

---

26 https://en.wikipedia.org/wiki/Occam%27s_razor

Which absurd auxiliary thesis will come next? Maybe the one by Dr. Richard Green, who seriously proposed that the cyanide residues in the disinfestation chambers did not stem from fumigations, but were caused by objects leaning against the wall which had been soaked in a "hydrogen-cyanide solution"? Where then do the cyanide residues close to the ceiling, in the middle and outside of the wall come from? And why did the plaster of the two above-mentioned churches turn blotchy blue all the way up to the ceiling?

Seen in the light of day, the intellectual capers and mental contortions of orthodox dogmatists are absurd to such an extent that they are indeed comical…

* * *

In the following table, in the first white group, samples from alleged homicidal gas chambers are listed. The second, gray group contains samples from Zyklon-B disinfestation chambers. The third group lists samples from locations that don't belong to either group.

Table 5: Analyses results of masonry samples, total cyanide

No.	Location (Auschwitz, if not stated differently)	Sample taker	mg/kg
1-7	Crematory II, Morgue #1 ("gas chamber")	Leuchter	0.0
8	Crematory III, Morgue #1 ("gas chamber")	Leuchter	1.9
9	Crematory III, Morgue #1 ("gas chamber")	Leuchter	6.7
10,11	Crematory III, Morgue #1 ("gas chamber")	Leuchter	0.0
13,14	Crematory IV, foundation wall	Leuchter	0.0
15	Crematory IV, foundation wall	Leuchter	2.3
16	Crematory IV, foundation wall	Leuchter	1.4
17-19	Crematory IV, foundation wall	Leuchter	0.0
20	Crematory IV, foundation wall	Leuchter	1.4
21	Crematory V, foundation wall	Leuchter	4.4
22	Crematory V, foundation wall	Leuchter	1.7
23,24	Crematory V, foundation wall	Leuchter	0.0
25	Crematory I, morgue ("gas chamber")	Leuchter	3.8
26	Crematory I, morgue ("gas chamber")	Leuchter	1.3
27	Crematory I, morgue ("gas chamber")	Leuchter	1.4
29	Crematory I, morgue ("gas chamber")	Leuchter	7.9
30	Crematory I, morgue ("gas chamber")	Leuchter	1.1
31	Crematory I, morgue ("gas chamber")	Leuchter	0.0
1	Crematory II, Morgue #1 ("gas chamber")	Rudolf	7.2
2	Crematory II, Morgue #1 ("gas chamber")	Rudolf	0.6
3	Crematory II, Morgue #1 ("gas chamber")	Rudolf	6.7/0.0
	Crematory II, Morgue #1 ("gas chamber")	Mattogno	0.0
	Crematory II, Morgue #1 ("gas chamber")	Mattogno	0.0
3	Crematory II, Morgue #1 ("gas chamber")	Ball	0.4
4	Crematory III, Morgue #1 ("gas chamber")	Ball	1.2
5	Bunker 2, foundation	Ball	0.1
6	Crematory V, foundation wall	Ball	0.1
32	Delousing chamber BW 5a, inside	Leuchter	1,050.0
9	Delousing chamber BW 5a, inside	Rudolf	11,000.0
11	Delousing chamber BW 5a, inside	Rudolf	2,640.0/1,430.0
12	Delousing chamber BW 5a, inside	Rudolf	2,900.0
13	Delousing chamber BW 5a, inside	Rudolf	3,000.0
14	Delousing chamber BW 5a, outside	Rudolf	1,035.0
15a	Delousing chamber BW 5a, outside	Rudolf	1,560.0
15c	Delousing chamber BW 5a, outside	Rudolf	2,400.0
16	Delousing chamber BW 5b, outside	Rudolf	10,000.0
17	Delousing chamber BW 5b, inside	Rudolf	13,500.0
18	As well as, BW 5a, wood from door frame	Rudolf	7,150.0
19a	Delousing chamber BW 5b, inside	Rudolf	1,860.0
19b	Delousing chamber BW 5b, inside	Rudolf	3,880.0
20	Delousing chamber BW 5a, inside	Rudolf	7,850.0
22	Delousing chamber BW 5a, inside	Rudolf	4,530.0
1	Delousing chamber BW 5b, inside and outside	Ball	3,170.0
2	Delousing chamber BW 5a, inside and outside	Ball	2,780.0
28	Crematory I, lavatory	Leuchter	1.3
	Crematory II, Morgue #2 ("undressing room")	Mattogno	1.2
	Crematory II, Morgue #2 ("undressing room")	Mattogno	1.3
5	Detainee barracks	Rudolf	0.6
6	Detainee barracks	Rudolf	<0.1
7	Detainee barracks	Rudolf	0.3
8	Detainee barracks	Rudolf	2.7/0.0
23	Detainee barracks	Rudolf	0.3
24	Detainee barracks	Rudolf	0.1
25	Brick of Bavarian farmhouse	Rudolf	9.6/9.6

Concentrations in mg cyanide (CN^-) per kg masonry material (brick, mortar, concrete, plaster). Cyanide values lower than 10 mg/kg are uncertain; samples with values below 1-2 mg are considered free of cyanide. When two values are shown, the second value represents the result of a control analysis conducted by another company.

The Cremations of Auschwitz

This chapter is intentionally not titled "Crematories of Auschwitz" or "The cremation furnaces of Auschwitz," because this is neither about the buildings where cremation furnaces were located, nor exclusively about cremations in these furnaces. As mentioned earlier, it is rather safe to say that in Auschwitz corpse cremations on outdoor pyres did indeed occur in the second half of 1942 as a consequence of the apocalyptic typhus epidemic which broke out in early 1942, and due to an inadequate or outright lack of cremation capacities that prevented conventional cremations in furnaces.

According to the legend, from May 1944 onward, such mass incinerations are said to have taken place again as a consequence of the alleged extermination of the Hungarian Jews.

In the following, we will therefore first deal with the cremation furnaces of Auschwitz, and after that with the open-air incineration pyres alleged for 1944.

The Cremation Furnaces

Flames

Here's a rumor we have to quash right up front: a vast number of witnesses related how gigantic flames burst from the chimneys of the crematories of Auschwitz. This would have been technically completely impossible. From the construction blueprints it can be seen that the flues of the largest

***Illustration 34:** Painting by former Auschwitz detainee David Olère of Crematory III in Birkenau – with flames and thick smoke from the oversized chimney (compare Illustration 39).*

crematories in Auschwitz (Nos. II & III), running from the furnaces to the chimney, had a length of 15 m (some 49 feet), and the chimney itself was also about 15 m tall. These furnaces were fueled by coke or coal, and in rare cases also by wood, especially when heating up after the necessary daily cleaning of the fireplace grates. Coke burns with an extremely short flame. Coal has a somewhat larger flame, but can hardly exceed half a meter (1.6 feet). Wood fires can indeed form longer flames, but in light of the fact that the fireplaces of the cremation furnaces were quite small, not a lot of wood fit in there.

In other words: the fuel could never have caused flames that could even protrude out of the cremation muffle. This is different in the case of fat that comes from a corpse lying in the muffle. When this is set free in sufficiently large quantities all at once, the developing fat vapors can, while burning, enter the smoke flue and, in case of sufficient quantities and a sufficiently strong draft, could indeed shoot out of the chimney. For this to happen, however, obese individuals are required – of which there were practically none in Auschwitz.[27]

The president of the Austrian Chamber of Engineers at that time, Walter Lüftl, has commented on this problem as follows in a technical magazine:[28]

> *"We know from past cases: Even if 46 witnesses more or less firmly declare that they heard nothing, the 47th witness who heard something, whose statement can be verified by experts, nonetheless speaks the truth.*
> *On the other hand, it is strange that in certain proceedings relating to cremation facilities, testimony perhaps is given that 'meter-high flames shot out of high chimneys,' although this is technically impossible, since as a rule only warm exhaust gases flow out of chimneys (except in quite rare explosions – with gas heating, perhaps) and there is never even a reflection to be seen, because the flames (as in the case of coke firing) are unable to leave the combustion chamber, and the reflection is dissipated in the flue."*

Consequently, the witnesses have lied and have, in a manner of speaking, "copied lies" from each other and from the ruling script – across the board, hundreds, even thousands of them.

Smoke

And what about the smoke that, according to the statements of mostly the same witnesses, constantly covered the whole camp, yes, the whole area? It's technically very difficult to utilize a coal or coke combustion in a smokeless way. It is to be expected that the crematories of Auschwitz indeed discharged a certain amount of smoke while in operation. The problem is, however, that in *all* of the aerial photos taken by Allied reconnaissance planes of the Birkenau Camp since late May 1944 not even from *one* crematory a column of smoke

---

27 See the experiments by Mattogno 2004a.
28 Lüftl 1991; quoted in Rudolf 2017b, pp. 193f.

Illustration 35: *The coal-hydrogenation plants of IG Farbenindustrie AG near Auschwitz-Monowitz in the winter of 1943/1944.*

can be seen (Rudolf 2018, Chapter 5). Back then, these facilities allegedly would have been cremating at their peak capacity, as at that time about 300,000 of the approximately 400,000 Jews deported from Hungary were allegedly killed and incinerated in Auschwitz. In the next section, I will go into this in more detail.

Auschwitz is situated in a bottomland, and the chemical industry that was erected at that time a few kilometers east of the camp, receiving its power from local coal power stations and which in diverse processes burned coal, respectively coke, will have discharged significantly more smoke than the Crematories of Auschwitz were ever capable of. Back then, such industrial facilities were not equipped with smoke filters. During high-pressure weather conditions with little wind, this smoke gathered in the river valley covering the whole area with stinking smog. This is the likely background of these eye- and nose-witness testimonies

Claimed Capacity

As noted, the Soviet and Polish investigation commissions organized their "findings" as well as their selected witnesses in such a manner that they supported the four-million lie. Accordingly, they also exaggerated the capacity of the Auschwitz crematory furnaces, to which they attributed absurd capabilities, incinerating up to 10,000, 20,000, 30,000, 40,000, yes, even more than 60,000 corpses per day (Mattogno 2015, pp. 311-320).

Today the Holocaust orthodoxy acknowledges these exaggerations as such, insists, however, that the capacity had been 4,800 corpses per day because these numbers were found in a document in the archives of the camp.[29]

What is the truth?

Actual Capacity

The coke-fired cremation furnaces of Auschwitz were manufactured by the company Topf & Söhne of the city of Erfurt.[30] Although the facilities installed in the various crematories of Auschwitz were different from each other in some of their details, all were built following the same pattern. Every muffle had a door with a width and height of 60 cm (1.97 feet), of which the upper half consisted of a semicircular arch. In comparison to civilian cremation furnaces, these were cheaper versions whose muffles and doors were reduced in size, as they were meant to only take in one normal corpse at a time *without* casket. Civilian cremation muffles, on the other hand, had to also take in obese corpses plus the casket. Unfortunately, hardly any documents about the operation of the Auschwitz crematories are left or have been

29 See Komitee… 1957, p. 269; Kogon *et al.* 1993, p. 157; Pressac 1989, p. 247; *Der Spiegel* No. 40/1993, p. 151; Bailer-Galanda *et al.* 1995, p. 69.

30 The observations in this chapter are based on the study by Mattogno/Deana 2015.

Illustration 36: *Coke-fired double-muffle furnace by the Topf Company in the Mauthausen Concentration Camp with identical muffles to those that were installed in Auschwitz. The two horizontal lines represent the height of two corpses on top of each other on the muffle grate (Mattogno 2015, p. 721).*

made available to date. Fortunately, such documents for a crematory that had a very similar furnace by the Topf company do exist: the crematorium in Gusen Concentration Camp in Austria that exists to this day. From the partially preserved documentation of this installation one can read that, per incineration of a normal corpse, approximately 30 kg (66 pounds) of coke was needed, and that the full incineration of a single corpse had a duration of about an hour. During that time, the corpse (respectively its parts) stayed in the muffle for around 40 to 45 minutes, after which it disintegrated and fell through the fireclay grate into the ashpit beneath it, where the cremation was completed.

The furnaces installed in Auschwitz had some construction differences compared to the furnace in Gusen, resulting in a somewhat-longer cremation duration:

Illustration 37: *Interior of the Topf Company furnace, type Auschwitz, with slits in the fireclay grate with a width of merely 1.97 inch.*

1. In Auschwitz the slits of the fireclay grates of the muffles had a width of only approximately 5 cm (1.97 inches; see Illustration 37), whereas the Gusen model had larger openings of 25 cm × 30 cm (9.84 × 11.8 inches; see Illustration 38). That means that corpse parts of much larger dimensions could fall through the grate a lot earlier in the Gusen furnace than in the Auschwitz type. So, if one wanted to insert the next corpse as soon as this became possible, while the parts of the previous corpse were still burning in the ashpit, in Gusen one could do so after 40 to 45 minutes, but in Auschwitz only markedly later. Such a procedure was illegal, however, because in doing this, the ashes of the corpses got mixed, but in the case of catastrophic epidemics as they at times raged in Auschwitz, these concerns would have surely been put aside – as well as in the hypothetical case of mass exterminations.
2. Moreover, none of the crematories in Birkenau was equipped with a forced-draft blower that served to enhance the draft of the chimney. Such a blower brought

Illustration 38: *Interior of the muffle of the Topf Company furnace, Type Gusen, with gaps of 11.81 inch (length) and 9.84 inch (width) in the fireclay grate.*

about a larger air current through the burning coke, hence a stronger heat development, similar to a bellows. In Auschwitz the heat supply to the muffles thus depended solely on the natural draft of the chimneys. This would have prolonged the cremation time as well at a concurrently slight reduction of coke usage per time unit.

Hence, all things considered, with the furnaces of Auschwitz it took about an hour before the next body could be inserted into the muffle. This rate, by the way, was confirmed by the responsible engineers of the Topf Company, who at the end of the war were apprehended in the city of Erfurt by the Soviets and interrogated in a quite rough manner (see Graf 2002). Detailed cremation experiments conducted in the 1920s in Switzerland, as well as extensive cremation-technical calculations confirm this rate.

If one considers that the fireplace had to be cleaned daily from ash and cinders which required extinguishing the fire and cooling down the furnace, then the needed time of firing

Illustration 39: *Crematory II being built, January 1943.*

Illustration 40: *Crematory III in the spring of 1943.*

Illustration 41: *Crematory IV being built, winter 1942/43.*

Illustration 42: *Crematory V in the summer of 1943.*

up the furnace, this means that a coke-fired crematory could be operated for a maximum of about 20 hours per day.

The Birkenau Crematories II and III each had five triple-muffle furnaces, thus a total of 30 muffles.

In both the Birkenau Crematories IV and V, four double-muffle furnaces were assembled into one block, so that each crematory had eight muffles, resulting in a total of 16.

Towards the end of its operational period, the old Crematory I in Auschwitz Main Camp had three double-muffle furnaces, so a total of six muffles.

For all crematories together, this results in (30+16+6=) 52 muffles. Crematory I, however, was shut down in the summer of 1943, and around the same time, Crematory IV got damaged beyond repair and was never fixed again. Crematory III only became operational when Crematory IV broke down, and in March 1943, Crematory II broke down again for several months shortly after having been put into operation, as the flues and the chimney got damaged due to irregular strain as well as overload.

Illustration 43: *The five triple-muffle furnaces in Crematory II. Photo taken in the beginning of 1943*

Thus, there effectively never were more than (30+8=) 38 cremation muffles concurrently operational in the Auschwitz Camp. Their maximum daily capacity at a 20-hour operation amounted to:

38 muffles × 20 hours × 1 corpse/hour = 760 corpses.

Please note that this is the theoretical maximum capacity when operating the furnaces in a competent and appropriate way. As they were operated by unskilled detainees, however, of whom one can presume an understandable tendency towards sabotage instead of care, one can assume that the actual capacity was markedly lower.

In addition, single furnaces or even complete crematories had to be shut down on occasion for necessary repairs, reducing the effective capacity even further (see Mattogno/Deana 1994).

That capacity only states how many corpses *could have been* incinerated. What do we know, however, about the number of corpses that were *actually* cremated?

Number of Corpses Cremated in the Crematories

A set of documents about the operation of the crematories of Auschwitz has been preserved: quantities of coke delivered there in the period from February 1942 to October 1943. The documentation about this is almost complete and results in the data compiled in Table 6 (Pressac 1989, p. 224):

Table 6: Monthly coke deliveries to the Auschwitz crematories in metric tons

Month '42	metric tons	Month '43	metric tons
February	22	January	23
March	39	February	40
April	39	March	144.5
May	32	April	60
June	25	May	95
July	16.5	June	61
August	31.5	July	67
September	52	August	71
October	15	September	61
November	17	October	82
December	39	**Total**	**1032,5**
∅ 2/42-2/43:	30	∅ 3/43-10/43:	80

From February 1942 to February 1943, only the old crematory in the Main Camp was in operation. Until the end of March 1942, it had two fully functional double-muffle furnaces, then three from April 1942 (Mattogno 2016a, p. 89). In this period, on average 30 metric tons of coke were used monthly. If one assumes on the basis of the similarity between these furnaces and those in Gusen that the cremation therein of a corpse required 30 kg (66 lbs) of coke, then this results in the numbers listed in Table 7 for monthly and daily cremations (assuming a work week of six days) as well as daily cremations per muffle.

If one considers that there are some gaps in the documents about coke deliveries, and that wood was also delivered to the crematories, these numbers would increase slight-

Table 7: Monthly and daily cremations in Crematory I (Main Camp) at 30 kg (66 lbs) of coke/corpse

Month	Corpses	Corpses/Day	Corpses/Muffle/Day
February '42	733	33.3	8.3
March '42	1,300	48.1	12.0
April '42	1,300	54.2	9.0
May '42	1,067	39.5	6.6
June '42	833	34.7	5.8
July '42	550	20.4	3.4
August '42	1,050	38.9	6.5
September '42	1,733	72.2	12.0
October '42	1,500	55.6	9.3
November '42	567	23.6	3.9
December '42	1,300	48.1	8.0
January '43	767	28.4	4.7
February '43	1,333	60.6	10.1
Total	**14,033**		

ly accordingly. If one additionally considers that the demand for coke per cremation at continuous operation slightly decreases – the costly heating up of a cooled-down furnace becomes unnecessary – then these numbers increase again by a certain percentage (on this, see Mattogno 2019).

Let us now compare these numbers to the numbers kept by the camp staff in the Auschwitz Death Books about regular deaths of registered detainees, see Illustration 16 (p. 32). In Table 8, the numbers of the previous table are compared to those of the Death Books.

If one corrects the possible monthly cremations on the basis of the above-mentioned error margin, it becomes clear that the crematory in the Main Camp, at the latest from May 1942 on, wouldn't have been able to cope with the requirements of the continuously increasing numbers of dead. Indeed, at the end of May it was discovered that the crematory's chimney showed cracks (Mattogno 2016a, pp. 48f.), which points to an overloading of the facilities to the breaking point.

Table 8: Comparison of monthly cremations calculated from the usage of coke to monthly deaths

Month	Cremations	Deaths	Load [%]
February '42	733	1,087	148
March '42	1,300	2,446	188
April '42	1,300	1,671	129
May '42	1,067	2,160	202
June '42	833	2,335	280
July '42	550	4,345	790
August '42	1,050	8,507	810
September '42	1,733	7,199	415
October '42	1,500	4,492	299
November '42	567	958	169
December '42	1,300	997	77
January '43	767	4,454	581
February '43	1,333	5,907	443

Despite growing mortality, the number of cremations decreased in June probably because starting in June construction work on a new chimney began. Starting in mid-July until early August, the old crematory had to be shut down completely because in this period the old chimney was dismantled and the new one connected to the flue. Coke deliveries were accordingly low, and with that the capacity for cremations during these months.

But exactly in July and August the typhus epidemic reached its peak. In these two months alone, a total of almost 13,000 detainees died, though according to Table 8 only about 1,600 of them could be cremated. An upward-adjusted number on the basis of not-counted wood and coke deliveries as well as economies due to continuous operation will not have been much higher than 2,000 either.

The excess of around 11,000 corpses, added to the excesses of the months of May and June (approximately 2,000) were buried in mass graves in Birkenau (see Illustration 17, p. 33), although they had to be exhumed later. From

Illustration 44: *Painting by former Auschwitz detainee David Olère of an incineration pit in Birkenau – babies are thrown alive from the truck into the flames. Although pyres most probably existed in the second half of 1942, Olère's interpretation is brimming with "artistic license".*

mid/end September 1942, all remains that could not be cremated were then most probably directly burned on pyres.

From March 1943 on, the new crematories were gradually brought into operation in Birkenau, whereupon Crematory II broke down shortly after and Crematory IV permanently, due to massive damage probably caused by a too-early raising of the temperature of the not-yet-set masonry, as mentioned earlier.

If one takes a look at the coke deliveries from March 1943 on, only March is eye-catching with extremely high quantities. In the following months the usage of coke settles at a quantity that was only around a factor of 2.5 higher than in the period before, when only the old crematory was in operation. Until the summer of 1943, not just the 6 muffles of the old crematorium were available, but – not counting the failing of Crematory IV – altogether (6 +15 +15 +8 =) 44 muffles, and after the shutdown of Crematory I from the autumn of 1943, 38 muffles. The cremation capacity thus was (44 ÷ 6 =) 7$^{1}/_{3}$ times, or from autumn 1942 on still (38 ÷ 6 =) 6$^{1}/_{3}$ times greater than before.

This means in plain language that the crematories of Birkenau weren't even half as intensively used as the old one was before. Even if one considers that the efficiency of the new furnaces was a bit better, that is to say the usage of coke per hour was somewhat less than with the old ones (by one-third for Crematories II & III, and half for Crematory V), it still is true that the new crematories by far weren't used as intensively as the old one at a time when it had to cope all by itself with the entire load of all corpses. In other words: The SS had provided for an enormous capacity that they didn't use at all in the end.

It's certain that the coke deliveries starting March 1943 approximately matched the numbers of the dead mentioned in the Death Books, if one assumes a realistic coke demand per corpse of around 30 kg (66 lbs) for Crematory I, 20 kg (44 lbs) for Crematories II & III, and about 15 kg (33 lbs) for Crematories IV and V.[31]

For additional hundreds of thousands of corpses that allegedly accrued according to the orthodox legend since March 1943 from alleged mass gassings, there simply was no fuel available. In order to uphold the legend anyway, orthodox scholars have the nerve to claim that for the incineration of a corpse only 3.5 kg (7.7 lbs) of coke was required (van Pelt 2002, p. 122).

Oh, sure, and pigs can fly…

---

31 An accurate calculation of the number of corpses that could be incinerated with the delivered quantities of coke from March 1943 on depends on the (unknown) distribution of the incinerations among all available crematories, on the duration of the operation times of each crematory, and on the efficiency of the furnace used. To unravel those here would lead too far. See Mattogno/Deana 1994; Mattogno 2015, pp. 301-305, 449-451; Mattogno 2019 for details.

The Durability of a Cremation Muffle

The fireproof grate and the fireproof lining of a cremation muffle do not last forever. The Topf Company, which had built the furnaces in Auschwitz and Birkenau, specified a durability for these components of about 3,000 incinerations, which was 50% above the standard in those days (Jakobskötter 1941, p. 583). The specification, a marketing fudge, thus may have been an exaggeration. If one considers that the crematories in Auschwitz and Birkenau were operated by an unskilled and hostile staff, namely detainees, one can imagine that this specification given by the Topf Company is an optimistic maximum

After over 3,000 incinerations, the fireclay grate and the muffle lining thus needed to be renewed. To do this, they had to disassemble and reassemble the furnace completely. That was an enormous task and would have spawned a multitude of documents in the archives of Auschwitz.

It's a fact, however, that there are no documents for any of the crematories in Auschwitz that indicate it had even been considered to carry out such an overhaul, l*et al*one that it had been performed.

If we turn a blind eye to reality and assume that all these muffles really had performed and withstood 3,000 incinerations, wherein we exclude the failing Crematory IV that broke down after a few cremations, we then arrive at a maximum number of (44 muffles × 3,000) = 132,000 cremations.

In the parts of the Death Books of Auschwitz that to date have been made available to the public, which with a few exceptions cover the years 1941 through 1943, just under 70,000 deaths are registered. These, please note, all pertain to properly registered detainees of the camp, who *did not* die in the gas chambers but in another way. Some volumes of these years are missing, and the years 1940 and 1944 as well as January 1945 are completely absent. If one extrapolates the available data to these periods, the resulting number may be around 100,000.

There is simply no room for an additionally million or so of gassed victims. For that, the furnaces would have needed to withstand, with just one fireclay lining, 30,000 incinerations. This is an utter impossibility. Alternatively, it would have been needed to renew this lining about ten times, which would have left extensive documentation and all kinds of witness recollections. Not a trace of this can be found.

Illustration 45-47: *Photos taken by the SS of the arrival of Hungarian Jews in Auschwitz, spring of 1944, taken from the* Auschwitz-Album *(see Klarsfeld).*

The Pyres

As was explained earlier, pyres must have existed in Auschwitz from late summer 1942 on, on which those thousands were incinerated who had fallen victim to typhus and whose bodies had to be exhumed from the intolerably polluting mass graves that were dug earlier, and those who died thereafter who could not be incinerated in the crematory of the Main Camp due to overload.

The orthodox Auschwitz narrative claims, however, that there were similar incinerations on pyres from mid-May 1944 until far into the summer of that year. Their background is the deportation of over 400,000 Jews from Hungary to Auschwitz that took place between the middle of May and early July 1944.

Extensive documentation with photos exists about these deportations, compiled by the SS at Auschwitz upon the arrival of these transports (see Illustrations 45-47). However, there are very few documents as to what happened to these

***Illustration 48:** Foundation walls of a building to the west of the Birkenau Camp, said to be of the so-called Bunker 2 allegedly used for mass gassings. © Carlo Mattogno, July 1992*

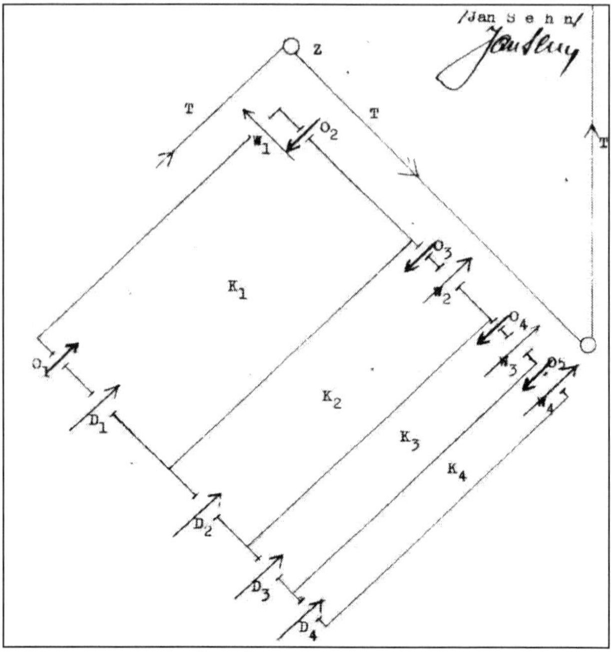

Illustration 49: *The floor plan of Bunker 2 claimed by key witness Shlomo Dragon for this building – recorded by the Polish lying judge Jan Sehn. As one can see from Illustration 48, it has nothing to do with reality (Mattogno 2016c, Document 12, p. 225).*

deported people after arriving in Auschwitz (see Mattogno 2001).

Orthodox historians claim that around three quarters of the deported Jews were immediately killed on arrival in the purported extermination facilities at Birkenau, meaning in the gas chambers. Because the cremation capacities of the crematories would have been too low for this enormous number of corpses, most of the killed victims are said to have been burned on pyres outdoors.

The pits wherein these pyres are said to have burned were allegedly located roughly north of Crematory V and in the wider vicinity of the so-called Bunker 2, west of the Birkenau Camp.

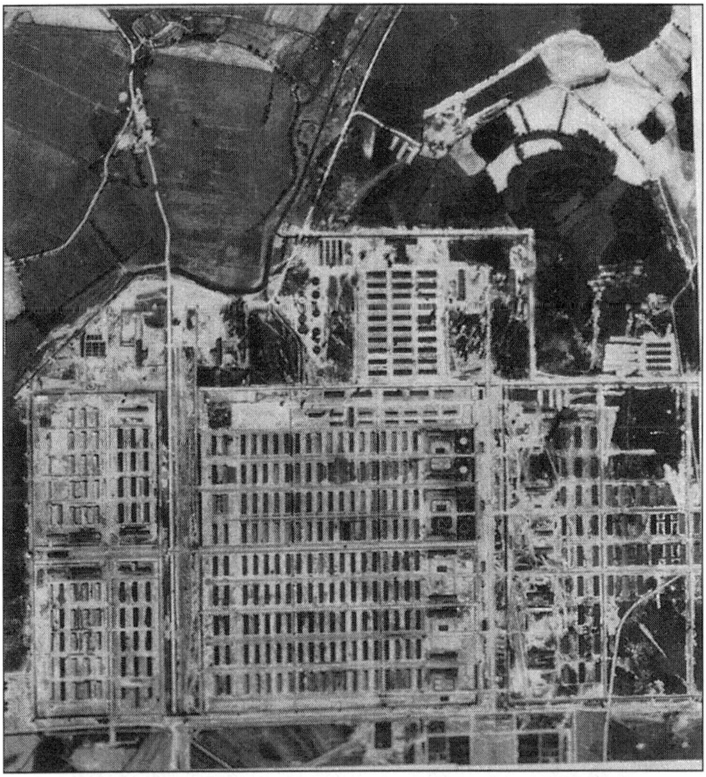

Illustration 50: *US aerial photo of May 31, 1944 of the Auschwitz-Birkenau Camp.*

"Bunker 2" supposedly was an old farmhouse that the SS is said to have converted into a makeshift gassing facility. For this claim there is neither documentary nor material support. In the area where this building is said to have been located, there are to date remainders of wall foundations, but neither the size nor the floorplan of these foundations bears any resemblance to what witnesses have claimed about them (see Illustrations 48f.).

Due to the complete absence of other evidence, the Auschwitz orthodoxy thus solely relies on stories of witnesses in this respect. The most prominent of these reports were given in the same propagandistic context and partly even by

***Illustration 51:** Detail enlargement of Illustration 50 of the area where the pyres are said to have been.*

the same witnesses who, amongst other things, also reported about flame-spouting crematory chimneys and insanely short cremation times. At that time, everything was just flatly aimed at supporting the mendacious four-million propaganda of the Soviets.[32]

Contrary to the evidence, we start our following deliberations with the premise that the alleged mass gassings in this Bunker 2 and the subsequent incinerations of the victims' bodies on pyres took place as claimed, indeed. What would

[32] As to the totally invented stories about the so-called Bunkers of Auschwitz, see Mattogno 2016c.

the logistics of such an operation have been? And what visible traces would this have left on the aerial photos that were taken by Allied reconnaissance aircraft around this period? Illustration 50 shows one of them. It was taken on May 31st 1944 by a US reconnaissance aircraft (Rudolf 2018, p. 55).

Illustration 52f.: *Photos of cadaver incinerations outdoors during a foot-and-mouth epidemic in England in 2001.* *(www.whale.to/m/fmd70.html)*

According to preserved documents, around 184,000 Jews had been deported from Hungary by that date. The first of them arrived in Auschwitz on May 17. If we assume that three quarters of these people were killed immediately after their arrival, this results in approximately 138,000 corpses that had to be disposed of.

As mentioned earlier, in the spring of 1944 only three crematories were operational. The old crematory in the Main Camp had been shut down in the summer of 1943, and Crematory IV in Birkenau had suffered irreparable damage right from the start, so was also out of commission.

Therefore 38 cremation muffles were available:

- Crematory II: 5 furnaces with 3 muffles each = 15 muffles
- Crematory III: 5 furnaces with 3 muffles each = 15 muffles
- Crematory V: 4 furnaces with 2 muffles each = 8 muffles

If we assume a realistic cremation capacity of one average adult per muffle per hour – with a maximum daily operation time of 20 hours – then this means that in the discussed fifteen days from May 17 until May 31 (15 days × 38 muffles

×20 corpses/muffle/day=) 11,400 corpses of adults could be incinerated in the crematories. If we include children, who burn faster, this amount might have increased by about 20%, which results in approximately 13,680, which we will round up to 14,000.

Illustration 54f.: *More photos of the foot-and-mouth epidemic in England. Incomplete cremations on pyres.*

Therefore, of the 138,000 allegedly murdered (138,000 – 14,000 =) 124,000 could not have been incinerated in the crematories but would have had to be burned outdoors on pyres. This results in a daily average of about 8,267. I'll round this number off to 8,000 because to a certain degree we are dealing here with estimates, and I do not want to give the impression that these numbers are exact. But this gives us a rather good approximation of the dimension of the problem to be solved.

The next question is, how to incinerate so many corpses every day on pyres outdoors?

Fortunately – or unfortunately, depending on how one looks at it – quite good empirical data exists as to the mass incinerations of corpses on pyres. These were gathered during the outbreak of a foot-and-mouth epidemic among the livestock of England in 2001.[33]

This tragic event caused the death of approximately six million cows, pigs and sheep – the similarity with another number is purely coincidental. Because the cremation capaci-

33 See https://en.wikipedia.org/wiki/2001_United_Kingdom_foot-and-mouth_outbreak

ties for offal in England were completely insufficient to process such amounts of animal cadavers, the authorities fell back on open-air incinerations.

Illustration 56: *The infernal heat of large pyres requires the use of thermal protective clothing.*

The documentation compiled during this epidemic was gathered by a German researcher who took the following parameters from it, which are significant for our case (Köchel 2004, 2016):

1. In order to operate the surface of a pyre, it should not be wider than 2.5 m (8.2 ft) and not higher than 2 m (6.6 ft), the latter also in order to prevent the heap from toppling over.
2. On average one needs about 135 kg (298 lbs) of wood to incinerate a cadaver of the weight of an average human being. This quantity is doubled when fresh wood is used, because it possesses only half the calorific value of dry wood.
3. From these constraints it results that the density of the heap of such a pyre, when using dry wood, would be about 8 to 10 average human corpses per running meter (3.3 ft), though only half of that when using freshly cut wood.
4. Pyres of these sizes generally burn a whole day long, but the heat of the embers under the upper ash layer continues for days, up to a week if the heap is left undisturbed.
5. The infernal heat generated by such pyres makes it impossible to stay in its vicinity for a prolonged period without protective clothing, *let alone* to work there. In case such pyres are located near to each other, one had to

leave a minimal distance of approximately 50 m (164 ft) between them in order to enable work at either of these pyres while the other was burning. This also means that sufficient space must be available for the transport of the bodies and the fuel, as well as for the removal of the ash and other burnt residue.

In the case examined here, witnesses claim that the ash was later sifted by detainees looking for remains not completely burned up, and that larger pieces were crushed manually. I won't discuss here to what extent this would have been possible for the cremation remains of 8,000 corpses per day. The fact is that incinerations on pyres are incomplete by their nature, so that there would have been a large quantity that had to be sifted and crushed.

We are only interested here in the fact that sifting and crushing of the remains would have been possible only after the pyre had significantly cooled down, so at the earliest after two days. This means that the capacity of the pyres had to be 16,000 corpses daily, because the burning up and processing of every pyre would have taken at least two days. And with this, the time required for the sifting and crushing of the remains has not even been considered.

If the Auschwitz camp authorities would have succeeded in procuring the required quantities of dry fuel wood at all times – so approximately (8,000 corpses × 135 kg/corpse = 1.080.000 kg =) 1,000 metric tons daily, for which not the slightest bit of documentation or anecdotal proof exists – a single pyre for 16,000 corpses would have been approximately (16,000 corpses ÷ 10 corpses/m =) 1.6 km (1 mile) long. In case freshly cut wood from the surrounding forests would have been used, the equivalent pyre would have been twice as long, hence 2 miles. Would they have split this pyre into, for instance, ten equally long, parallel pyres instead with a distance of 50 m (164 ft) from each other in order to be able to be tended concurrently, then these pyres would have

Table 9: Claimed and required numbers of pyres for the problem in question (Mattogno 2016b, p. 28)

Witness	Length	Claimed number	Required number
F. Müller	50 m	5	32
D. Paisikovic	50 m	2	32
M. Nyiszli	50 m	2	32
H. Mandelbaum	35 m	several	46
C. Mordowicz/A. Rosin	30 m	several	53
S. Dragon	25 m	5	64
A. Feinsilber/Jankowski	20 m	2	80
S. Bendel	12 m	3	133
J. Rosenblum	10 m	several	160

taken up a surface area of approximately (160 m × (10 × 2.5 m + 9 × 50 m) =) 76,000 m² (818,000 sq ft or some 19 acres). This amount is then again doubled in case of the use of fresh wood.

Let's now have a look at what the most-important witnesses have claimed about these pyres. Table 9 shows the name of the witness in the left column, next to that the length of the pyre claimed by the witness, then the claimed number of pyres, and lastly the number of pyres of this length that would have been needed to incinerate 8,000 bodies daily with dry wood. This number needs to be doubled for fresh wood.

This makes it clear that none of the witnesses had the slightest idea of the dimensions of the logistical problems the claimed incinerations outdoors would have entailed. They "solved" the problem mostly by stating absurd stacking densities and heights while understating the required quantity of fuel or fully ignoring it. In that way, witness Henryk Mandelbaum claimed for instance that the pyres had consisted of ten alternating layers of corpses and fuel wood resulting in a pyre of at least around 10 m high (33 ft, cf. Mattogno 2016b, p. 19).

Illustration 57: *Livestock-cadaver incineration in England in 2001 seen from the air. The alleged significantly larger burning locations in Auschwitz would have covered the whole area in smoke. As a matter of fact, nothing of that can be seen in the Allied aerial photos.*

Thus, what should we see in the aforementioned aerial photos? Well, smoke; a lot of smoke. Just like the livestock-cadaver incinerations in England of 2001 caused enormous quantities of smoke, see Illustration 57.

Now have another look at the aerial photo of May 31, 1944 (pp. 77f.). Do you see any smoke there? As already mentioned, not even the chimneys of the crematory show smoke.

Illustration 58 shows the same aerial photo, but with smoke added by means of Photoshop that *should* be there.

Compare both images and see the difference.

Especially this peaceful aerial photo is irrefutable proof that the whole story of the extermination of the Hungarian Jews and their incineration on enormous pyres outdoors has been nothing other than a gigantic propaganda lie. There's no changing that.[34]

---

[34] For more revelations of the lie of mass murder of the Hungarian Jews see Mattogno 2001.

Illustration 58: *Detail enlargement of the aerial photo of May 31, 1944 with* **applied** *smoke as it would have had to look if the witnesses were telling the truth: massive formation of smoke from the area top left (Bunker 2) and north of Crematory V (center lower half).*

Establishment Reactions

British historian David Irving was so much taken with Fred Leuchter's expert report submitted during the second Zündel Trial that he agreed to testify for the defendant as an expert historian in this trial (see Lenski 1990, starting on p. 399). A year later he even published a glossy edition of the Leuchter Report with his own preface.

The Holocaust lobby never forgave Irving for this and subsequently did everything possible to destroy his career. This campaign was so successful that Irving lost practically all his book contracts. Cornered that way, he decided to sue a protagonist of these lobbyists, Dr. Deborah Lipstadt, and her publishing company for libel. This litigation, backfired, however, when he lost the lawsuit decidedly.[35]

Illustration 59: *David Irving*

In the German-speaking realm, the revisionist seed fell on very fertile ground due to the patriotic bump in the course of the German reunification of 1989/90. A row of prominent intellectuals showed a lot of interest and were open to taking the

35 Cause of Irving's wrath was Lipstadt's 1993 book *Denying the Holocaust: The Growing Assault on Truth and Memory*; on this book, see Rudolf 2017c; as to the trial, see Guttenplan 2001; Lipstadt 2005; Mattogno 2015.

scientific results of revisionism seriously (as described in Rudolf 2017b, pp. 133-139).

Illustration 60:
Prof. Dr. Hellmut Diwald

The most prominent of those was the already mentioned public involvement of the then president of the Austrian Federal Chamber of Civil Engineers, Walter Lüftl (see Rademacher 2003). The most far-reaching, though, was probably the curiosity that Leuchter awoke in the young chemist Germar Rudolf, because unperturbed by threats and violent measures against him, he made it his life's work to put the complete Holocaust issue onto a basis of exact scholarship with radical thoroughness (see Rudolf 2016a&c).

Rudolf's chemical expert report about Auschwitz was only the first step. Here are the voices of several German historians on this expert report, set down in private letters to the author:

> *"I am extraordinarily impressed. To my knowledge, you are the first expert in Germany who has addressed this particular topic in a scholarly impeccable and well-founded way. It is not for me to attribute an ice-breaker function to your expert report. It is easy to see which political-historical effects will originate from it, though its entire dimension cannot yet be estimated."*

Prof. Dr. H. Diwald, Historian, January 22, 1992

> *"I read it with great interest.* […] *My impression is, however, that this expert report is an important contribution to a very important question which, since the 'Leuchter Report,' needs to be answered urgently.* […] *One can only very much hope that the well-known tactics of hushing up is not applied to your expert report, but that critical responses and comments will be made."*

Prof. Dr. E. Nolte, Historian, January 1, 1992

"I rank the receipt of your study to be one of the climaxes of knowledge one is able to experience in our times. Together with quite a few colleagues in the area of contemporary history, I share joy and gratitude with regard to the research work you took upon yourself, and of course especially in view of the results of your sound scientific examination."

Illustration 61:
Prof. Dr. Ernst Nolte

Prof. Dr. W.G. Haverbeck, Historian, January 31, 1992

"I took my time to read your study! It raises hope to see that a representative of the younger generation without bias, with scientific thoroughness and noticeable great expertise and corresponding researcher inquisitiveness, in a worldwide-contested question, sets off to get to the bottom of it! Clear and distinct the result! In the long run true circumstances cannot be suppressed! I wish your work to achieve the breakthrough!"

Prof. Dr. E. Schlee, Historian, April 01, 1992

"[…] *I would fervently wish that all statements about this set of problems would be so manifestly based on long and intensive labor as yours is. Doubtlessly most cannot be scrutinized by the layman but the photos as such are already very informative."*

Prof. Dr. E. Nolte, Historian, January 01, 1993

Shortly afterwards, Rudolf's expert report was also published in Dutch and French. On April 7, 1995, for a broadcasted feature of the political magazine "Panorama" of the Belgian TV station BRT 1, Hans Westra of the Dutch Anne Frank Foundation was questioned about Rudolf's expert report. Westra expounded:

"Rudolf is a young scientist who, in an excellently laid out work with tables, charts and so on, tries to prove that the gas chambers were technically impossible [...] *These scientific analyses are perfect."*

Illustration 62: *Hans Westra*

In 1997, the Swiss publisher René-Louis Berclaz was put on trial by the Jewish League against Racism and Anti-Semitism (Ligue Internationale contre le Racisme et l'Antisémitisme, LICRA) for, amongst other things, the dissemination of the French edition of Rudolf's report, because it allegedly incites to "racial discrimination." Berclaz filed a motion that the court obtain expert opinions as to whether Rudolf's report is scientific and therefore allowed to be freely distributed.

Judge Jean-Pierre Schröter of the 3rd District Court in Châtel-Saint-Denis, charged to hear the case, granted the motion and ordered, amongst others, the chemist and certified court expert Prof. Dr. Henri Ramuz to review Rudolf's report to the effect of whether it is scientific in nature and if its conclusions were correct. On May 18, 1997, Dr. Ramuz delivered his expert opinion on Rudolf's report to the court by stating:

"In the course of last week, I have read many passages of the Rudolf Report. I did this as a chemist who takes note of the experiments of another chemist. As you have seen, the Rudolf expert report consists of five parts:

a. Description of the local and miscellaneous facilities of the concentration camp (Auschwitz-Birkenau).

b. Examination of the physico-chemical properties of hydrogen cyanide, its chemical reactivity in general, but also the actions of iron with formation of blue pigments (Prussian blue).

c. *Examination of the reactivity of hydrogen-cyanide gas on living organisms and comparison of the operation techniques for the gassing of apparel (pest control) and gassings of human beings; connected to this is the experiment to demonstrate that the first technique was applied successfully, while the application of the second technique is seen by the author as practically impossible.* [36]

Illustration 63:
The current edition of Rudolf's expert report.

d. *Results of analyses of samples that, according to the author, stem from the gas chambers as well as from the delousing chambers. These analyses were conducted by various individuals as well as by various institutes.*

e. *General conclusions.*

In my answer I can only tackle some elements of the second part of this report which is about the chemistry of hydrogen-cyanide gas and some of its derivatives. Overall, he bases himself on literature that was compiled long before the report and has to be designated as scientifically correct.

Concerning the physiochemical interaction with components of the masonry and the stability of these molecular compounds in the course of several decades, this is an extremely complex subject: The influence of meteorological factors, of rain, of sun radiation, erosion by wind and the large differences in temperatures (between -40 and + 140 degrees Fahrenheit) as well as lastly the complex and variable configuration of the chambers exposed to these factors, does not permit me to make a statement about the conclusions of the author of

36 Rudolf only regards as impossible the kind of homicidal gassings *as described by the witnesses*, not gassings of humans in general.

this expert report. A scientific result can only be said to be proven if several researchers under the same reaction conditions and by means of the same samples, procured from the assured origin, obtain the same results.[37]

In the scientific realm, Germar Rudolf is no amateur; he understands inorganic chemistry, analytic chemistry and physical chemistry. The way he has procured the samples, by whom he had them analyzed,[38] *how he, as a person familiar with science, interprets them, to all of this I cannot make a statement nor issue an expertise. All German top people in the field of inorganic chemistry have received this expert report. It would be better to write: 'There was no cause for comments.'*

Hoping that this expert opinion may be of some use to you, I present to you, dear sir (judge),
my cordial greetings.

Henri Ramuz
May 18, 1997

As a consequence of this expert report, René-Louis Berclaz was not prosecuted for the dissemination of Rudolf's report, because the report was scientific.

Still: Because that cannot exist which is not allowed to exist, the Holocaust orthodoxy defended itself the only way it could react to such convincing research results. In one European country after the other, they succeeded to put the governments under pressure – or the brains of the rulers have been softened to such an extent by life-long Holocaust propaganda that no pressure was required – that increasingly restrictive censorship laws were passed, to which everyone

37 Why are Rudolf's research results not fully verified, something Rudolf had always demanded, when to this date it hasn't been possible to find any mistake made by him?

38 The samples taken from Auschwitz and other locations were examined by the renowned German analysis institute, Institut Fresenius. This has been proven in all detail in Rudolf's publications. Why then this peculiar objection?

falls victim who expresses an undesired view of the Holocaust:

1990: France prohibits Holocaust revisionism
1992: Austria prohibits Holocaust revisionism
1994: Germany prohibits Holocaust revisionism
1995: Switzerland prohibits Holocaust revisionism
1995: Belgium prohibits Holocaust revisionism
1997: Luxemburg prohibits Holocaust revisionism
1998: Poland prohibits Holocaust revisionism
2000: Liechtenstein prohibits Holocaust revisionism
2001: Slovakia prohibits Holocaust revisionism
2001: Czechoslovakia prohibits Holocaust revisionism
2002: Romania prohibits Holocaust revisionism
2005: Germany tightens its prohibition
2007: The EU recommends prohibition to all members
2007: Portugal prohibits Holocaust revisionism
2010: Hungary prohibits Holocaust revisionism
2012: Lithuania prohibits Holocaust revisionism
2014: Russia prohibits Holocaust revisionism
2014: Greece prohibits Holocaust revisionism
2016: Italy prohibits Holocaust revisionism*
2017: The UK prohibits Holocaust revisionism*

* Only when done in conjunction with mocking or defaming the victims.

In Israel Holocaust revisionism is prohibited since 1986, and in Canada and Australia one can be charged by "Human Rights Commissions" under civil law…

In Germany, it is moreover prohibited by threat of punishment to file a motion with the court to admit evidence seeking to prove that revisionist statements are correct. Reason: such motions themselves commit the offense of "denial" and would therefore represent a criminal act during a

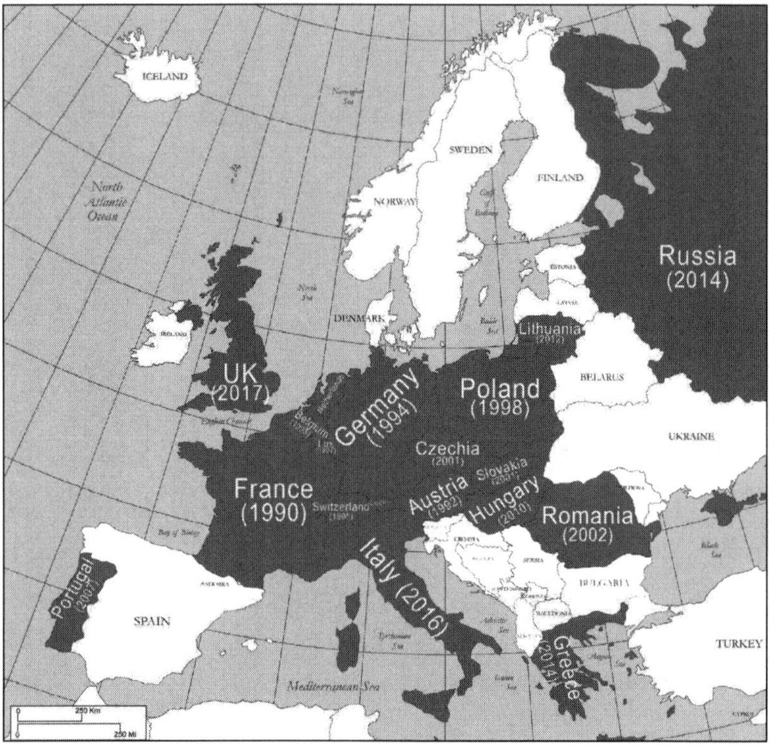

Illustration 64: *Censorship map of Europe: All dark countries threaten to put peaceful historical dissidents in the dungeon …*

public trial…[39] Several defense lawyers have been sentenced accordingly, simply because they filed such a motion.

However, the denial of all motions to introduce evidence and the prosecution of lawyers who are not deterred by this is still not the end of German judicial repression in such cases. Certain valiant defense lawyers would not let themselves be gagged and, despite threats by the prosecutor and the judge, bravely kept filing motions to introduce evidence by which they attempted to defend their clients. The result of

[39] 5th Criminal Division of the German Federal High Court, StR 5 StR 485/01; see *Neue Juristische Wochenschrift* 2002, p. 2115; *Neue Strafrechts-Zeitung*, 2002, p. 539; Martin 2002.

that was the passing of the gag paragraph §257a into the German code of criminal procedure that enables judges to order defense lawyers to file any contribution to the trial in writing only, except for their final plea. And that's exactly what is happening on a regular basis in such cases.

In order to keep the public in the dark about the fact that the defendants are being sent to the dungeon for entirely harmless and scientifically well-based statements, their writings for which they are on trial are moreover not read out in the courtroom, which would normally be obligatory, but the trial participants – judges, prosecution, defense – are ordered to read the material by themselves at home.

Ever since the introduction of these measures, silence has been once more every citizen's primary duty in German courtrooms. Shut up, and don't you dare protest!

At the end of such a show trial, during which the defense is almost completely paralyzed, a verdict follows into which the judges can write almost anything they want. Hardly anything can be checked anyway, because during German penal trials, no record is kept of what is being said. This way, the judges boost their career, are submissive to the lynch media and bow down to politics.[40]

Ever since, there's been graveyard peace.

Our thoughts are free, our thinkers are imprisoned.

---

40 For more details on the sad dictatorial state of the German judicial system watch Rudolf's documentary "Germany, Country under the Rule of Law: Role Model or Illusion?", Aug. 27, 2017; https://codoh.com/library/document/4872/.

Resistance Is Obligatory!

When in 2009 the German Federal Constitutional Court had to determine if the sections of the German penal code used to prosecute historical dissidents (§130 of the Penal Code) are in breach of the German constitutional principles of freedom of opinion, its decision was revealing:[41]

> *"In general, restrictions to the freedom of opinion are permissible only on the basis of general laws according to art. 5, para. 2, alternative 1, Basic Law. A law restricting opinions is an inadmissible special law, if it is not formulated in a sufficiently open way and is directed right from the start only against certain convictions, attitudes or ideologies.* […] *Although the regulation of art. 130, para. 4, German Penal Code is not a general law* […] *even as a non-general law it is still compatible with art. 5, para. 1 and 2, Basic Law, as an exception. In view of the injustice and the terror caused by the National-Socialist regime, an exception to the prohibition of special laws* […] *is immanent."*

Or in other words: exceptional laws are prohibited, except in exceptional cases. Here the logic of this exception goes as follows:

> *Because Germany in the past, in breach of the Weimar Constitution, burned books and persecuted and locked up peaceful dissidents, Germany today is morally obliged, in breach of the Bonn Constitution, to burn books and persecute and lock up peaceful dissidents.*

---

41 BVerfG, 1 BvR 2150/08, Nov. 4, .2009; see www.bundesverfassungsgericht.de/pressemitteilungen/bvg09-129.html

Illustration 65: *Courtroom of the German Federal Constitutional Court.*

The fact is that §130 of the German penal code, already at its invention in 1849(!),[42] aimed "from the start only against certain convictions, attitudes or ideologies" and has not lost that property to this day. It is therefore clearly unconstitutional from the beginning to the end.

No judicial system in the world needs criminal laws to prohibit the expression of certain opinions. If somebody misused his right of expression of opinion to publicly call for the violation of the civil and human rights of others, then in all judicial systems this is already covered by the prohibition of inciting to a crime (§26 German Penal Code) or the prohibition of publicly calling for the committing of criminal acts (§111 German Penal Code). Only these laws deserve the name of "general laws." Any additional law of censorship is nothing but the spawn of dictatorial thought which every German has the right and the moral obligation to resist as soon as the German Federal Constitutional Court denies remedy, so states Article 20 Chapter 4 of the current German constitution.

---

42 Prussia introduced this law to prosecute communist and socialist agitators as well as other antimonarchic groups.

Illustration 66:
Joschka Fischer

George Orwell, in his original preface to his book *Animal Farm* that was rejected by four publishing houses under pressure of the British Government, amongst others, expressed it as follows:[43]

> *"If liberty means anything at all, it means the right to tell people what they do not want to hear."*

The quality of a judicial system is marked by its capability to protect those members of the population against persecution by the State who harbor opinions which especially the powerful wish to suppress. This is primarily about those people who break the central taboos of a society or erode its founding myths, hence those whose critique goes to the roots of a society. As long as these opinions are peaceful, thus do not call for the violation of rights of others or justify such a violation, the judicial system is *prohibited* to penalize the public expression of such opinions.

But what are the central taboos and the foundational myths of current German society and also very important cornerstones of many other Western societies?

In 1999, the then German Minister of Foreign Affairs Joschka Fischer stated the following (Lévy 1999):

> *"All democracies have a basis, a foundation. For France it's 1789. For the US the Declaration of Independence. For Spain the Spanish Civil War. Well, for Germany it's Auschwitz. It can only be Auschwitz. The remembrance of Auschwitz, the 'never again*

[43] See http://orwell.ru/library/novels/Animal_Farm/english/efp_go.

Auschwitz,' in my eyes can be the only foundation of the new Berlin Republic."

One could quote a long list of personalities and media voices expressing what they think if someone shakes this foundation. We don't have to do this here, though, because everyone knows what the vast majority of the population in Germany and most of the Western World thinks of those of whom it is rumored that they "deny" Auschwitz and/or the Holocaust as a whole. For many, such people are at a similar moral level as child molesters. One could hardly sink any lower.

What would *you*, dear reader, do, if someone came to you and told you in a peaceful and possibly even scientific-objective way something about Auschwitz you would never want to hear? That's almost a rhetorical question in a society in which almost a monolithic consensus exists as to what to do with such taboo-breaking historical dissidents:

"Lock them up! Hang them higher!"

But exactly here the crucial question arises: What do you think of the rule of law? Can and will the legal framework of your country protect peaceful dissidents of the history of the Third Reich from assaults by the State and society, or will they be thrown to the wolves?

The tough reality of today's Germany unfortunately demonstrates that the German judicial system in particular is well-nigh tailor-made to enforce political expectations with judicial penal power without restraint.[44]

One of the most-brutal enforcers against historical dissidents, the already-mentioned German judge Ulrich Meinerzhagen of the Mannheim District Court was quoted by the leftist newspaper *tageszeitung* as follows in relation to

---

44 For details, see the documentary movie *Germany, Country under the Rule of Law: Role Model or Illusion?* By Germar Rudolf, https://codoh.com/library/document/4872/

the trial of historical dissident Ernst Zündel (Klingelschmidt 2007):

> *"In the end the court rejected all motions with the terse – and for some anti-fascists in the audience shocking – reasoning that it was completely irrelevant whether the Holocaust had happened or not. Its denial is punishable in Germany. And only that counts before the court. 'Democracy must be able to endure that,' a law student later lectured in the foyer of the court building."*

Because as we all know, democracy is when three foxes and a chicken decide what to have for supper – or here, that the overwhelming majority of all members of a society under threat of punishment can instruct which opinion *you*, dear reader, may publicly express about certain historical topics and which not.

Apparently, the law student has not understood that the constitutional state was exactly invented in order prevent such assaults of the majority on minorities.

But in the end, you, valued spectator, probably need not to be concerned. Because when the judiciary came for the Holocaust deniers, you could confidently remain quiet; after all, you were not a Holocaust denier. When they imprisoned the Nazis, you were again silent; after all, you were not a Nazi. When they came to fetch the right wingers, you still kept silent, as you weren't a right winger either. When they then one day will come for you, nobody will be left to protest.

Good luck then with such a mockery of a judiciary!

For in Germany, everyone is evidently fair game!

* * *

Here is the wording of the concerned article on free speech in the German constitution:

> *"(1) Everyone has the right to freely express and disseminate his opinion verbally, in writing and in image, and without hindrance to inform himself through publicly available source. The freedom of*

press and the freedom of reporting by radio and film are guaranteed. Censorship does not take place.
(2) These rights have their limits in the rules of the general laws, the statutory regulations concerning the protection of youth and in the right to personal honor."

Illustration 67:
Prof. Dr. Roman Herzog

It says that censorship does not take place, but Germany's judiciary thinks this only means that there is no censorship *before* something is stated publicly. After the publication of a disfavored opinion, censorship is merrily enforced in Germany! After all, in Paragraph 2 above, freedom of opinion is immediately abrogated again, because if according to the German Federal Constitutional Court non-general censorship laws also are just fine, then no freedom of expression exists.

In 1970, a German professor of public law, who at that time taught at the University for Administrative Science in Speyer, wrote in an obscure *Festschrift* the following trenchant words on the right of German citizens to resist assaults by the State on the human rights warranted in the German constitution (quoted acc. to Peters 2005, p. 184):

> *"Seen in the cold light of day, every single section of the Constitution is… nothing more than the concrete elaboration of one of these basic principles of Western constitutional statehood, so that the attack against practically every single section simultaneously touches upon the basic principles of Section 20 of the Constitution* [the Germans' right to resist]*."*

17 years later, the author of these lines was elected president of the German Federal Constitutional Court, and seven years

after that even to be the German federal president: Professor Dr. Roman Herzog (Herzog 1970, p. 100). The complete evisceration of the freedom of expression in Germany took place during his terms in office.

* * *

In summary:

1. The judicial system of the Federal Republic of Germany is medieval in its structure and repressive in its adjudication.
2. Against such systematic violations of the human rights by the rulers, to which no other remedy is possible, all Germans – and not just they – have the right and the moral obligation to resist:

Illustration 68:
Frederick the Great

> *"Against anyone who undertakes to undermine this order, all Germans have the right of resistance, if other remedy is not possible."*
>
> Section 20, Paragraph 4 of the Constitution of the Federal Republic of Germany

* * *

> *"You need to know that the least of peasants, and what is even more, the beggar is just as much a human being as is his majesty, and he has to find justice by that fact that all humans are equal before the law; it may be a prince suing the peasant or vice versa, then the prince will be equal to the peasant before the law: and in such affairs it has to be proceeded purely by justice with no regard to the person. The justice councils in all provinces have to only comply with this. And wherever they do not go straight forward with justice without regard of person or class and put aside natural justness, they*

shall get in trouble with his royal majesty. A legal council which exercises injustices is more dangerous and worse than a gang of thieves; one can protect oneself against those, but nobody can protect himself against rogues who use the robes of justice to carry out their vicious passions; they are worse than the biggest scoundrels in the world and deserve double punishment."

Frederick the Great
(quoted acc. to Frank 1926, p. 99)

* * *

"So long as the superstition that men should obey unjust laws exists, so long will their slavery exist."

Mahatma Gandhi
(Narayan 1969, p. 174)

Illustration 69:
Mahatma Gandhi

Bibliography

Note: Major booksellers (such as Amazon and Barnes & Noble) have removed many of the revisionist titles listed below from their offerings. Such titles, however, can be obtained from Castle Hill Publishers, including in downloadable form at no charge, at the URLs listed for each.

- Alvarez, Santiago, *The Gas Vans: A Critical Investigation*, Castle Hill Publishers, Uckfield 2016 (©2011); https://shop.codoh.com/book/57
- Auswärtiges Amt (ed.), *Amtliches Material zum Massenmord von Katyn*, F. Eher, Berlin 1943; https://archive.org/details/Auswaertiges-Amt-Amtliches-Material-zum-Massenmord-von-Katyn
- Bailer-Galanda, Brigitte, Wolfgang Benz, Wolfgang Neugebauer (eds.), *Wahrheit und Auschwitzlüge*, Deuticke, Vienna 1995
- Christianson, Scott, *The Last Gasp: The Rise and Fall of the American Gas Chamber*, University of California Press, Berkeley, Cal., 2010
- Dokumentationszentrum des österreichischen Widerstandes, Bundesministerium für Unterricht und Kultur (eds.), *Amoklauf gegen die Wirklichkeit*, Vienna 1991.
- Frank, Bruno, *Friedrich der Große als Mensch im Spiegel seiner Briefe*, Deutsche Buch-Gemeinschaft, Berlin 1926
- Fritz Bauer Institut, Staatliches Museum Auschwitz (eds.), *Der Auschwitz-Prozess: Tonbandmitschnitte, Protokolle, Dokumente*, (DVD) Directmedia Publishing, Berlin 2004
- Lüftl, Walter, "A Somewhat Different Auschwitz Trial," *The Revisionist*, Vol. 2, No. 3 (2004), pp. 294f.; https://codoh.com/library/document/1712/
- Gärtner, Michael; Werner Rademacher (= Willy Wallwey), "Grundwasser im Gelände des KGL Birkenau (Auschwitz)," *Vierteljahreshefte für freie Geschichtsforschung*, Vol. 2, No. 1 (1998), pp. 2-12; http://vho.org/VffG/1998/1/GaeRad1.html
- Gauss, Ernst 1993 (=Germar Rudolf), *Vorlesungen über Zeitgeschichte: Strittige Fragen im Kreuzverhör*, Grabert, Tübingen 1993
- Gauss, Ernst 1994 (ed.), *Grundlagen zur Zeitgeschichte: Ein Handbuch über strittige Fragen des 20. Jahrhunderts*, Grabert, Tübingen 1994
- Graf, Jürgen, "Anatomie der sowjetischen Befragung der Topf-Ingenieure: Die Verhöre von Fritz Sander, Kurt Prüfer, Karl Schultze und Gustav Braun durch Offiziere der sowjetischen Antispionageorganisation Smersch (1946-1948)," *Vierteljahreshefte für freie Geschichtsforschung*, Vol. 6, No. 4 (2002), pp. 398-421; http://vho.org/VffG/2002/4/Graf398-421.html

- Graf, Jürgen, Carlo Mattogno, *Concentration Camp Majdanek: A Historical and Technical Study*, 3rd ed., Castle Hill Publishers, Uckfield 2016 https://shop.codoh.com/book/96
- Green, Richard J., "Report of Richard J. Green," introduced into evidence at Queen's Bench Division, Royal Courts of Justice, Strand, London, *David John Cawdell Irving v. (1) Penguin Books Limited, (2) Deborah E. Lipstadt*, ref. 1996 I. No. 1113, 2001; http://phdn.org/archives/holocaust-history.org/irving-david/rudolf/affweb.pdf
- Guttenplan, Don D., *The Holocaust on Trial*, Goldmann, Norton, New York 2001
- Herzog, Roman, "Das positive Widerstandsrecht" in: *Festschrift für A. Merkel*, Munich 1970
- Hoffman, Michael A., *The Great Holocaust Trial*, 3rd ed., Wiswell Ruffin House, Dresden, N.Y., 1995
- Irmscher, Richard, "Nochmals: 'Die Einsatzfähigkeit der Blausäure bei tiefen Temperaturen,'" *Zeitschrift für hygienische Zoologie und Schädlingsbekämpfung*, Vol. 34 (1942), pp. 35f.
- Jakobskötter, Rudolf. "Die Entwicklung der elektrischen Einäscherung bis zu dem neuen elektrisch beheizten Heißlufteinäscherungsofen in Erfurt," *Gesundheits-Ingenieur*, Vol. 64, No. 43 (1941), pp. 579-587
- Kalthoff, Jürgen; Martin Werner, *Die Händler des Zyklon B*, VSA-Verlag, Hamburg 1998
- Klarsfeld, Serge, *The Auschwitz Album*, Beate Klarsfeld Foundation, New York 1978
- Klingelschmidt, Klaus-Peter, "Prozessposse vor dem Ende," *Die tageszeitung*, Feb. 9, 2007, p. 6; www.taz.de/!318416/
- Köchel, Heinrich 2004, "Leichenverbrennungen im Freien," *Vierteljahreshefte für freie Geschichtsforschung*, Vol. 8, No. 4 (2004), pp. 427-432; English: "Outdoor Incineration of Livestock Carcasses," *Inconvenient History*, Vol. 7, No. 1 (2015); http://inconvenienthistory.com/7/1/3361
- Köchel, Heinrich 2016, "Outdoor Incineration of Livestock Carcasses," in: Mattogno 2016b, pp. 128-140
- Kogon, Eugen; Hermann Langbein; Adalbert Rückerl *et al.* (eds.), *Nazi Mass Murder: A Documentary History of the Use of Poison Gas*, Yale University Press, New Haven 1993
- Komitee der antifaschistischen Widerstandskämpfer in der DDR (ed.). *SS im Einsatz*, Kongress-Verlag, Berlin 1957
- Kulaszka, Barbara (ed.), *Did Six Million Really Die? Report on the Evidence in the Canadian "False News" Trial of Ernst Zündel – 1988*, Samisdat Publishers Ltd., Toronto 1992; https://archive.org/details/DidSixMillionReallyDie_528
- Lambrecht, Wolfgang, "Zyklon B – eine Ergänzung," *Vierteljahreshefte für freie Geschichtsforschung*, Vol. 1, No. 1 (1997), pp. 2-5; http://vho.org/VffG/1997/1/Lambrecht1.html

- Lamker, Hans, "Die Kurzwellen-Entlausungsanlagen in Auschwitz, Teil 2," *Vierteljahreshefte für freie Geschichtsforschung*, Vol. 2, No. 4 (1998), pp. 261-272; http://vho.org/VffG/1998/4/Lamker4.html
- Lenski, Robert, *The Holocaust on Trial*, Reporter Press, Decatur, Ala., 1990.
- Leuchter, Fred A.; Robert Faurisson; Germar Rudolf, *The Leuchter Reports: Critical Edition*, 5th ed., Castle Hill Publishers, Uckfield 2017; https://shop.codoh.com/book/404
- Lévy, Bernard-Henri, "Ein paar Versuche, in Deutschland spazieren zu gehen," Interview mit Josef Fischer, *Frankfurter Allgemeine Zeitung*, Feb. 18, 1999, p. 46
- Lipstadt, Deborah E. 1993, *Denying the Holocaust: The Growing Assault on Truth and Memory*, The Free Press, New York, 1993
- Lipstadt, Deborah E. 2005, *History on Trial: My Day in Court with David Irving*, Ecco, New York 2005
- Loitfellner, Sabine 2002, *Die Rezeption von Geschworenengerichtsprozessen wegen NS-Verbrechen in ausgewählten österreichischen Zeitungen 1956 – 1975*, undated [2002]; www.nachkriegsjustiz.at/prozesse/geschworeneng/rezeption.pdf
- Loitfellner, Sabine 2006, "Auschwitz-Verfahren in Österreich. Hintergründe und Ursachen eines Scheiterns," in: Thomas Albrich, Winfried Garscha, Martin Polaschek (eds.), *Holocaust und Kriegsverbrechen vor Gericht: Der Fall Österreich*, StudienVerlag, Innsbruck/Vienna/Bozen 2006
- Lüftl, Walter, "Sachverständigenbeweis versus Zeugenbeweis," *Konstruktiv*, No. 166 (1991), pp. 31f.
- Markiewicz, Jan 1994; Wojciech Gubala; Jerzy Labedz, "A Study of the Cyanide Compounds Content in the Walls of the Gas Chambers in the Former Auschwitz and Birkenau Concentration Camps," *Z Zagadnień Nauk Sądowych*, Vol. XXX (1994) pp. 17-27; http://codoh.com/library/document/4188/
- Martin, Sigmund P., "Volksverhetzung – Leugnen des Holocaust durch Verteidigerhandeln," *Juristische Schulung*, No. 11/2002, pp. 1127f.
- Mattogno, Carlo 2001, "Die Deportation ungarischer Juden von Mai bis Juli 1944. Eine provisorische Bilanz," *Vierteljahreshefte für freie Geschichtsforschung*, Vol. 5, No. 4 (2001), pp. 381-395; http://vho.org/VffG/2001/4/Mattogno381-395.html; English: "The Deportation of Hungarian Jews from May to July 1944: A Preliminary Account," https://codoh.com/library/document/357/
- Mattogno, Carlo 2002, "'Verbrennungsgruben' und Grundwasserstand in Birkenau," *Vierteljahreshefte für freie Geschichtsforschung*, Vol. 6, No. 4 (2002), pp. 421-428; http://vho.org/VffG/2002/4/Mattogno421-424.html; English: "'Cremation Pits' and Ground Water Levels at Birkenau," *The Revisionist*, Vol. 1, No. 1 (2003), pp. 13-16; https://codoh.com/library/document/1426/
- Mattogno, Carlo 2003a, "The Four Million Figure of Auschwitz: Origin, Revisions and Consequences," *The Revisionist*, Vol. 1. No. 4 (2003), pp. 387-392, 393-399; https://codoh.com/library/document/1535/

- Mattogno, Carlo 2004a, "Combustion Experiments with Flesh and Animal Fat," *The Revisionist*, Vol. 2, No. 1 (2004), pp. 64-72; https://codoh.com/library/document/1661/
- Mattogno, Carlo 2004b, "The Morgues of the Crematoria at Birkenau in the Light of Documents," *The Revisionist*, Vol. 2, No. 3 (2004), pp. 271-294; https://codoh.com/library/document/1713/
- Mattogno, Carlo 2005, *Auschwitz: Open Air Incinerations*, Castle Hill Publishers, Hastings 2005
- Mattogno, Carlo 2015, *The Real Case for Auschwitz: Robert van Pelt's Evidence from the Irving Trial Critically Reviewed*, Castle Hill Publishers, Uckfield 2015; https://shop.codoh.com/book/386
- Mattogno, Carlo 2016a, *Auschwitz: Crematorium I and the Alleged Homicidal Gassings*, 2nd ed., Castle Hill Publishers, Uckfield 2017; https://shop.codoh.com/book/110
- Mattogno, Carlo 2016b, *Auschwitz: Open-Air Incinerations*, 2nd ed., Castle Hill Publishers, Uckfield 2016; https://shop.codoh.com/book/477
- Mattogno, Carlo 2016c, *Debunking the Bunkers of Auschwitz: Black Propaganda versus History*, 2nd ed., Castle Hill Publishers, Uckfield, 2016; https://shop.codoh.com/book/12
- Mattogno, Carlo 2018, *The Einsatzgruppen in the Occupied Eastern Territories: Genesis, Missions and Actions*, Castle Hill Publishers, Uckfield 2018; https://shop.codoh.com/book/483
- Mattogno, Carlo 2019, *Auschwitz: Deliveries of Coke, Wood and Zyklon B: Neither Proof nor Trace for the Holocaust*, Castle Hill Publishers, Uckfield 2019 (in preparation); http://holocausthandbooks.com/index.php?page_id=40
- Mattogno, Carlo; Franco Deana 1994, "Die Krematoriumsöfen von Auschwitz-Birkenau," in: Gauss 1994, pp. 281-320; English updated as "The Crematoria Ovens of Auschwitz and Birkenau," in: Rudolf 2003, pp. 373-412
- Mattogno, Carlo; Franco Deana 2015, *The Cremation Furnaces of Auschwitz: A Technical and Historical Study*, Castle Hill Publishers, Uckfield 2015; https://shop.codoh.com/book/394
- Mazal, Harry W., "Zyklon-B: A Brief Report on the Physical Structure and Composition," (undated; 1998) http://phdn.org/archives/holocaust-history.org/auschwitz/zyklonb/
- McCarthy, Jamie; Mark van Alstine, "Zyklon Introduction Columns," undated; http://phdn.org/archives/holocaust-history.org/auschwitz/intro-columns
- Meeussen, Johannes C.L.; Erwin J.M. Temminghoff; Meindert G. Keizer; Ivo Novozamsky, "Spectrophotometric Determination of Total Cyanide, Iron Cyanide Complexes, Free Cyanide and Thiocyanate in Water by a Continuous-Flow System," *The Analyst*, Vol. 114 (1989), pp. 959-964
- Morris, Errol. *Mr. Death: The Rise and Fall of Fred A. Leuchter, Jr.*, Fourth Floor Productions, May 12, 1999; VHS: Universal Studios 2001; DVD: Lions Gate Home Entertainment, 2003; first shown on January 27,

1999 at the Sundance Film Festivals in Park City (Utah); https://youtu.be/YOqhuDGCC04
- Narayan, Shriman (ed.), *The Selected Works of Mahatma Gandhi*, Vol. 4, Navajivan Publishing House, Ahmedabad 1969
- Nowak, Hans Jürgen (=Willy Wallwey), "Kurzwellen-Entlausungsanlagen in Auschwitz," *Vierteljahreshefte für freie Geschichtsforschung*, Vol. 2, No. 2 (1998), pp. 87-105; http://vho.org/VffG/1998/2/Nowak2.html
- Nowak, Hans Jürgen, Werner Rademacher (=Willy Wallwey), "Some Details of the Central Construction Office of Auschwitz," in: Rudolf 2003, pp. 311-336
- Peters, Klaus, *Widerstandsrecht und humanitäre Intervention,* Osnabrücker Rechtswissenschaftliche Abhandlungen, Vol. 61, Carl Heymanns Verlag, Cologne 2005 (Dissertation der Univ. Osnabrück 2004/2005)
- Pressac, Jean-Claude 1989, *Auschwitz: Technique and Operation of the Gas Chambers*, Beate Klarsfeld Foundation, New York 1989; www.historiography-project.com/books/pressac-auschwitz/
- Pressac, Jean-Claude 1993, *Les crématoires d'Auschwitz: La machinerie du meurtre de masse*, CNRS, Paris 1993
- Pressac, Jean-Claude 1994, *Die Krematorien von Auschwitz: Die Technik des Massenmordes*, Piper, Munich 1994
- Rademacher, Werner, "The Case of Walter Lüftl: Contemporary History and the Justice System," in: Rudolf 2003, pp. 61-84
- Rudolf, Germar 2003 (ed.), *Dissecting the Holocaust: The Growing Critique of "Truth" and "Memory,"* 2nd ed., Theses & Dissertations Press, Chicago, Ill., 2003; https://shop.codoh.com/book/38
- Rudolf, Germar 2016a, *Resistance Is Obligatory: Address to the Mannheim District Court, 15 November 2006 to 29 January 2007*, 2nd ed., Castle Hill Publishers, Uckfield 2016; https://shop.codoh.com/book/414
- Rudolf, Germar 2016b, *Auschwitz: Plain Facts. A Response to Jean-Claude Pressac*, 2nd ed., Castle Hill Publishers, Uckfield 2016; https://shop.codoh.com/book/412
- Rudolf, Germar 2016c, *Hunting Germar Rudolf: Essays on a Modern-Day Witch Hunt*, Castle Hill Publishers, Uckfield 2016 https://shop.codoh.com/book/433
- Rudolf, Germar 2017a, *The Chemistry of Auschwitz: The Technology and Toxicology of Zyklon B and the Gas Chambers. A Crime-Scene Investigation*, 3rd ed., Castle Hill Publishers, Uckfield 2017; https://shop.codoh.com/book/436
- Rudolf, Germar 2017b, *Lectures on the Holocaust: Controversial Issues Cross-Examined*, 3rd ed., Castle Hill Publishers, Uckfield 2017; https://shop.codoh.com/book/446
- Rudolf, Germar 2017c, *Bungled: "Denying the Holocaust": How Deborah Lipstadt Botched Her Attempt to Demonstrate the Growing Assault on Truth and Memory*, 2nd ed., Castle Hill Publishers, Uckfield 2017; https://shop.codoh.com/book/443

- Rudolf, Germar 2017d, "Germany, Country under the Rule of Law: Role Model or Illusion?", Aug. 27, 2017; https://codoh.com/library/document/4872/
- Rudolf, Germar 2018 (ed.), *Air-Photo Evidence: World-War-Two Photos of Alleged Mass-Murder Sites Analyzed*, 5th ed., Castle Hill Publishers, Uckfield 2018; https://shop.codoh.com/book/451
- Sagel-Grande, Irene 1979; H.H. Fuchs; Christiaan F. Rüter (eds.), *Justiz und NS-Verbrechen*, Vol. 21, University Press, Amsterdam 1979
- Schulze, Dietmar, *Pulver und Schüttgüter: Fließeigenschaften und Handhabung*, Springer-Verlag, Berlin 2006
- Schwarz, L.; Walter Deckert 1929, "Experimentelle Untersuchungen bei Blausäureausgasungen," *Zeitschrift für Hygiene und Infektionskrankheiten*, Vol. 109 (1929), pp. 201-212
- Staatliches Museum Auschwitz-Birkenau 1995 (ed.), *Die Sterbebücher von Auschwitz*, Saur, Munich 1995
- Wellers, Georges 1991, "Der Leuchter-Bericht über die Gaskammern von Auschwitz," *Dachauer Hefte*, Vol. 7, No. 7 (November 1991), pp. 230-241
- Zimmermann Günter (ed.). *Bauschäden Sammlung*, Vol. 4, Forum-Verlag, Stuttgart 1981

Index of Names

Page numbers of entries from footnotes are in italics.

Free Samples at www.HolocaustHandbooks.com

HOLOCAUST HANDBOOKS

This ambitious, growing series addresses various aspects of the "Holocaust" of the WWII era. Most of them are based on decades of research from archives all over the world. They are heavily referenced. In contrast to most other works on this issue, the tomes of this series approach its topic with profound academic scrutiny and a critical attitude. Any Holocaust researcher ignoring this series will remain oblivious to some of the most important research in the field. These books are designed to both convince the common reader as well as academics. The following books have appeared so far, or are about to be released. Compare hardcopy and eBook prices at www.findbookprices.com.

SECTION ONE:
General Overviews of the Holocaust

The First Holocaust. The Surprising Origin of the Six-Million Figure. By Don Heddesheimer. This compact but substantive study documents

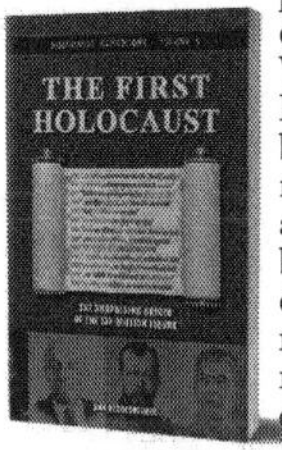

propaganda spread prior to, during and after the FIRST World War that claimed East European Jewry was on the brink of annihilation. The magic number of suffering and dying Jews was 6 million back then as well. The book details how these Jewish fund-raising operations in America raised vast sums in the name of feeding suffering Polish and Russian Jews but actually funneled much of the money to Zionist and Communist groups. 5th ed., 200 pages, b&w illustrations, bibliography, index. (#6)

Lectures on the Holocaust. Controversial Issues Cross Examined. By Germar Rudolf. This book first explains why "the Holocaust" is an important topic, and that it is essential to keep an open mind about it. It then tells how

many mainstream scholars expressed doubts and subsequently fell from grace. Next, the physical traces and documents about the various claimed crime scenes and murder weapons are discussed. After that, the reliability of witness testimony is examined. Finally, the author argues for a free exchange of ideas on this topic. This book gives the most-comprehensive and up-to-date overview of the critical research into the Holocaust. With its dialogue style, it is easy to read, and it can even be used as an encyclopedic compendium. 3rd ed., 596 pages, b&w illustrations, bibliography, index.(#15)

Breaking the Spell. The Holocaust, Myth & Reality. By Nicholas Kollerstrom. In 1941, British Intelligence analysts cracked the German "Enigma" code. Hence, in 1942 and 1943, encrypted radio communications between German concentration camps and the Berlin headquarters were decrypted. The intercepted data

Pictured above are all of the scientific studies that comprise the series *Holocaust Handbooks* published thus far or are about to be released. More volumes and new editions are constantly in the works. Check www.HolocaustHandbooks.com for updates.

refutes the orthodox "Holocaust" narrative. It reveals that the Germans were desperate to reduce the death rate in their labor camps, which was caused by catastrophic typhus epidemics. Dr. Kollerstrom, a science historian, has taken these intercepts and a wide array of mostly unchallenged corroborating evidence to show that "witness statements" supporting the human gas chamber narrative clearly clash with the available scientific data. Kollerstrom concludes that

the history of the Nazi "Holocaust" has been written by the victors with ulterior motives. It is distorted, exaggerated and largely wrong. With a foreword by Prof. Dr. James Fetzer. 5th ed., 282 pages, b&w ill., bibl., index. (#31)

Debating the Holocaust. A New Look at Both Sides. By Thomas Dalton. Mainstream historians insist that there cannot be, may not be, any debate about the Holocaust. But ignoring it does not make this controversy go away. Traditional scholars admit that there was neither a budget, a plan, nor an order for the Holocaust; that the key camps have all but vanished, and so have any human remains; that material and unequivocal documentary evidence is absent; and that there are serious problems with survivor testimonies. Dalton juxtaposes the traditional Holocaust narrative with revisionist challenges and then analyzes the mainstream's responses to them. He reveals the weaknesses of both sides, while declaring revisionism the winner of the current state

ISSN 1529-7748 · All books are 6"×9" unless otherwise stated. Artwork for paperback editions; plain hardcover editions available.

of the debate. 4th ed., 342 pages, b&w illustrations, bibliography, index. (#32)

The Hoax of the Twentieth Century. The Case against the Presumed Extermination of European Jewry. By Arthur R. Butz. The first writer to analyze the entire Holocaust complex in a precise scientific manner. This book exhibits the overwhelming force of arguments accumulated by the mid-1970s. Butz's two main arguments are: 1. All major entities hostile to Germany must have known what was happening to the Jews under German authority. They acted during the war as if no mass slaughter was occurring. 2. All the evidence adduced to prove any mass slaughter has a dual interpretation, while only the innocuous one can be proven to be correct. This book continues to be a major historical reference work, frequently cited by prominent personalities. This edition has numerous supplements with new information gathered over the last 35 years. 4th ed., 524 pages, b&w illustrations, bibliography, index. (#7)

Dissecting the Holocaust. The Growing Critique of 'Truth' and 'Memory.' Edited by Germar Rudolf. *Dissecting the Holocaust* applies state-of-the-art scientific techniques and classic methods of detection to investigate the alleged murder of millions of Jews by Germans during World War II. In 22 contributions—each of some 30 pages—the 17 authors dissect generally accepted paradigms of the "Holocaust." It reads as excitingly as a crime novel: so many lies, forgeries and deceptions by politicians, historians and scientists are proven. This is the intellectual adventure of the 21st Century. Be part of it! 3rd ed., 635 pages, b&w illustrations, bibliography, index. (#1)

The Dissolution of Eastern European Jewry. By Walter N. Sanning. Six Million Jews died in the Holocaust. Sanning did not take that number at face value, but thoroughly explored European population developments and shifts mainly caused by emigration as well as deportations and evacuations conducted by both Nazis and the Soviets, among other things. The book is based mainly on Jewish, Zionist and mainstream sources. It concludes that a sizeable share of the Jews found missing during local censuses after the Second World War, which were so far counted as "Holocaust victims," had either emigrated (mainly to Israel or the U.S.) or had been deported by Stalin to Siberian labor camps. 2nd ed., foreword by A.R. Butz, epilogue by Germar Rudolf containing important updates; 224 pages, b&w illustrations, bibliography (#29).

Air-Photo Evidence: World-War-Two Photos of Alleged Mass-Murder Sites Analyzed. By Germar Rudolf (editor). During World War Two both German and Allied reconnaissance aircraft took countless air photos of places of tactical and strategic interest in Europe. These photos are prime evidence for the investigation of the Holocaust. Air photos of locations like Auschwitz, Majdanek, Treblinka, Babi Yar etc. permit an insight into what did or did not happen there. The author has unearthed many pertinent photos and has thoroughly analyzed them. This book is full of air-photo reproductions and schematic drawings explaining them. According to the author, these images refute many of the atrocity claims made by witnesses in connection with events in the German sphere of influence. 6th edition; with a contribution by Carlo Mattogno. 167 pages, 8.5"×11", b&w illustrations, bibliography, index (#27).

The Leuchter Reports: Critical Edition. By Fred Leuchter, Robert Faurisson and Germar Rudolf. Between 1988 and 1991, U.S. expert on execution technologies Fred Leuchter wrote four reports on whether the Third Reich operated homicidal gas chambers. The first on Auschwitz and Majdanek became world-famous. Based on various arguments, Leuchter concluded that the locations investigated could never have been "utilized or seriously considered to function as execution gas chambers." The second report deals with gas-chamber claims for the camps Dachau, Mauthausen and Hartheim, while the third reviews design criteria and operation procedures of execution gas chambers in the U.S. The fourth report reviews Pressac's 1989 tome about Auschwitz. 4th ed., 252 pages, b&w illustrations. (#16)

Bungled: "The Destruction of the European Jews". Raul Hilberg's Failure to Prove National-Socialist "Killing Centers." By Carlo Mattogno. Raul Hilberg's magnum opus *The Destruction of the European Jews* is an orthodox standard work on the Holocaust. But how does Hilberg support his thesis that Jews were murdered *en masse*? He rips documents out of their context, distorts their content, misinterprets their meaning, and ignores entire archives. He only refers to "useful" witnesses, quotes fragments out of context, and conceals the fact that his witnesses are lying through their teeth. Lies and deceits permeate Hilberg's book, 302 pages, bibliography, index. (#3)

Jewish Emigration from the Third Reich. By Ingrid Weckert. Current historical writings about the Third Reich claim state it was difficult for Jews to flee from Nazi persecution. The truth is that Jewish emigration was welcomed by the German authorities. Emigration was not some kind of wild flight, but rather a lawfully determined and regulated matter. Weckert's booklet elucidates the emigration process in law and policy. She shows that German and Jewish authorities worked closely together. Jews interested in emigrating received detailed advice and offers of help from both sides. 2nd ed., 130 pages, index. (#12)

Inside the Gas Chambers: The Extermination of Mainstream Holocaust Historiography. By Carlo Mattogno. Neither increased media propaganda or political pressure nor judicial persecution can stifle revisionism. Hence, in early 2011, the Holocaust Orthodoxy published a 400-page book (in German) claiming to refute "revisionist propaganda," trying again to prove "once and for all" that there were homicidal gas chambers at the camps of Dachau, Natzweiler, Sachsenhausen, Mauthausen, Ravensbrück, Neuengamme, Stutthof... you name them. Mattogno shows with his detailed analysis of this work of propaganda that mainstream Holocaust hagiography is beating around the bush rather than addressing revisionist research results. He exposes their myths, distortions and lies. 2nd ed., 280 pages, b&w illustrations, bibliography, index. (#25)

SECTION TWO:
Specific non-Auschwitz Studies

Treblinka: Extermination Camp or Transit Camp? By Carlo Mattogno and Jürgen Graf. It is alleged that at Treblinka in East Poland between 700,000 and 3,000,000 persons were murdered in 1942 and 1943. The weapons used were said to have been stationary and/or mobile gas chambers, fast-acting or slow-acting poison gas, unslaked lime, superheated steam, electricity, Diesel-exhaust fumes etc. Holocaust historians alleged that bodies were piled as high as multi-storied buildings and burned without a trace, using little or no fuel at all. Graf and Mattogno have now analyzed the origins, logic and technical feasibility of the official version of Treblinka. On the basis of numerous documents they reveal Treblinka's true identity as a mere transit camp. 3rd ed., 384 pages, b&w illustrations, bibliography, index. (#8)

Belzec: Propaganda, Testimonies, Archeological Research and History. By Carlo Mattogno. Witnesses report that between 600,000 and 3 million Jews were murdered in the Belzec Camp, located in Poland. Various murder weapons are claimed to have been used: Diesel-exhaust gas; unslaked lime in trains; high voltage; vacuum chambers; etc. The corpses were incinerated on huge pyres without leaving a trace. For those who know the stories about Treblinka this sounds familiar. Thus the author has restricted this study to the aspects which are new compared to Treblinka. In contrast to Treblinka, forensic drillings and excavations were performed at Belzec, the results of which are critically reviewed. 142 pages, b&w illustrations, bibliography, index. (#9)

Sobibor: Holocaust Propaganda and Reality. By Jürgen Graf, Thomas Kues and Carlo Mattogno. Between 25,000 and 2 million Jews are said to have been killed in gas chambers in the Sobibór camp in Poland. The corpses were allegedly buried in mass graves and later incinerated on pyres. This book investigates these claims and shows that they are based on the selective use of contradictory eyewitness testimony. Archeological surveys of the camp are analyzed that started in 2000-2001 and carried on until 2018. The book also documents the general National-Socialist policy toward Jews, which never included a genocidal "final solution." 2nd ed., 456 pages, b&w illustrations, bibliography, index. (#19)

The "Operation Reinhardt" Camps Treblinka, Sobibór, Bełżec. By Carlo Mattogno. As an update and upgrade to the Volumes 8, 9 and 19 of this series, this study has its first focus on witness testimonies recorded during the World War II and the immediate post-war era, many of them discussed here for the first time, thus demonstrating how the myth of the "extermination camps" was created. The second part of this book brings us up to speed with the various archeological efforts made by mainstream scholars in their attempt to prove that the myth based on testimonies is true. The third part compares the findings of the second part with what we ought to expect, and reveals the chasm that exists between archeologically proven facts and mythological requirements. 402 pages, illustrations, bibliography, index. (#28)

Chelmno: A Camp in History & Propaganda. By Carlo Mattogno. At Chelmno, huge masses of Jewish prisoners are said to have been gassed in "gas vans" or shot (claims vary from 10,000 to 1.3 million victims). This study covers the subject from every angle, undermining the orthodox claims about the camp with an overwhelmingly effective body of evidence. Eyewitness statements, gas wagons as extermination weapons, forensics reports and excavations, German documents—all come under Mattogno's scrutiny. Here are the uncensored facts about Chelmno, not the propaganda. 2nd ed., 188 pages, indexed, illustrated, bibliography. (#23)

The Gas Vans: A Critical Investigation. By Santiago Alvarez and Pierre Marais. It is alleged that the Nazis used mobile gas chambers to exterminate 700,000 people. Up until 2011, no thorough monograph had appeared on the topic. Santiago Alvarez has remedied the situation. Are witness statements believable? Are documents genuine? Where are the murder weapons? Could they have operated as claimed? Where are the corpses? In order to get to the truth of the matter, Alvarez has scrutinized all known wartime documents and photos about this topic; he has analyzed a huge amount of witness statements as published in the literature and as presented in more than 30 trials held over the decades in Germany, Poland and Israel; and he has examined the claims made in the pertinent mainstream literature. The result of his research is mind-boggling. Note: This book and Mattogno's book on Chelmno were edited in parallel to make sure they are consistent and not repetitive. 398 pages, b&w illustrations, bibliography, index. (#26)

The Einsatzgruppen in the Occupied Eastern Territories: Genesis, Missions and Actions. By C. Mattogno. Before invading the Soviet Union, the German authorities set up special units meant to secure the area behind the German front. Orthodox historians claim that these units called *Einsatzgruppen* primarily engaged in rounding up and mass-murdering Jews. This study sheds a critical light onto this topic by reviewing all the pertinent sources as well as material traces. It reveals on the one hand that original war-time documents do not fully support the orthodox genocidal narrative, and on the other that most post-"liberation" sources such as testimonies and forensic reports are steeped in Soviet atrocity propaganda and are thus utterly unreliable. In addition, material traces of the claimed massacres are rare due to an attitude of collusion by governments and Jewish lobby groups. 2nd ed.., 2 vols., 864 pp., b&w illustrations, bibliography, index. (#39)

Concentration Camp Majdanek. A Historical and Technical Study. By Carlo Mattogno and Jürgen Graf. At war's end, the Soviets claimed that up to two million Jews were murdered at the Majdanek Camp in seven gas chambers. Over the decades, however, the Majdanek Museum reduced the death toll three times to currently 78,000, and admitted that there were "only" two gas chambers. By exhaustively researching primary sources, the authors expertly dissect and repudiate the myth of homicidal gas chambers at that camp. They also critically investigated the legend of mass executions of Jews in tank trenches and prove it groundless. Again they have produced a standard work of methodical investigation which authentic historiography cannot ignore. 3rd ed., 358 pages, b&w illustrations, bibliography, index. (#5)

Concentration Camp Stutthof and Its Function in National Socialist Jewish Policy. By Carlo Mattogno and Jürgen Graf. Orthodox historians claim that the Stutthof Camp served as a "makeshift" extermination camp in 1944. Based mainly on archival resources, this study thoroughly debunks this view and shows that Stutthof was in fact a center for the organization of German forced labor toward the end of World War II. 4th ed., 170 pages, b&w illustrations, bibliography, index. (#4)

SECTION THREE:
Auschwitz Studies

The Making of the Auschwitz Myth: Auschwitz in British Intercepts, Polish Underground Reports and Postwar Testimonies (1941-1947). By Carlo Mattogno. Using messages sent by the Polish underground to London, SS radio messages sent to and from Auschwitz that were intercepted and decrypted by the British, and a plethora of witness statements made during the war and in the immediate postwar period, the author shows how exactly the myth of mass murder in Auschwitz gas chambers was created, and how it was turned subsequently into "history" by intellectually corrupt scholars who cherry-picked claims that fit into their agenda and ignored or actively covered up literally thousands of lies of "witnesses" to make their narrative look credible. 2nd edition, 514 pp., b&w illustrations, bibliography, index. (#41)

The Real Case of Auschwitz: Robert van Pelt's Evidence from the Irving Trial Critically Reviewed. By Carlo Mattogno. Prof. Robert van Pelt is considered one of the best mainstream experts on Auschwitz. He became famous when appearing as an expert during the London libel trial of David Irving against Deborah Lipstadt. From it resulted a book titled *The Case for Auschwitz*, in which van Pelt laid out his case for the existence of homicidal gas chambers at that camp. This book is a scholarly response to Prof. van Pelt—and Jean-Claude Pressac, upon whose books van Pelt's study is largely based. Mattogno lists all the evidence van Pelt adduces, and shows one by one that van Pelt misrepresented and misinterpreted every single one of them. This is a book of prime political and scholarly importance to those looking for the truth about Auschwitz. 3rd ed., 692 pages, b&w illustrations, glossary, bibliography, index. (#22)

Auschwitz: Plain Facts: A Response to Jean-Claude Pressac. Edited by Germar Rudolf, with contributions by Serge Thion, Robert Faurisson and Carlo Mattogno. French pharmacist Jean-Claude Pressac tried to refute revisionist findings with the "technical" method. For this he was praised by the mainstream, and they proclaimed victory over the "revisionists." In his book, Pressac's works and claims are shown to be unscientific in nature, as he never substantiates what he claims, and historically false, because he systematically misrepresents, misinterprets and misunderstands German wartime documents. 2nd ed., 226 pages, b&w illustrations, glossary bibliography, index. (#14)

Auschwitz: Technique and Operation of the Gas Chambers: An Introduction and Update. By Germar Rudolf. Pressac's 1989 oversize book of the same title was a trail blazer. Its many document reproductions are still valuable, but after decades of additional research, Pressac's annotations are outdated. This book summarizes the most pertinent research results on Auschwitz gained during the past 30 years. With many references to Pressac's epic tome, it serves as an update and correction to it, whether you own an original hard copy of it, read it online, borrow it from a library, purchase a reprint, or are just interested in such a summary in general. 144 pages, b&w illustrations, bibliography. (#42)

The Chemistry of Auschwitz: The Technology and Toxicology of Zyklon B and the Gas Chambers – A Crime-Scene Investigation. By Germar Rudolf. This study documents forensic research on Auschwitz, where material traces and their interpretation reign supreme. Most of the claimed crime scenes – the claimed homicidal gas chambers – are still accessible to forensic examination to some degree. This book addresses questions such as: How were these gas chambers configured? How did they operate? In addition, the infamous Zyklon B can also be examined. What exactly was it? How does it kill? Does it leave traces in masonry that can be found still today? The author also discusses in depth similar forensic research conducted by other scholars. 4th ed., 454 pages, more than 120 color and over 100 b&w illustrations, bibliography, index. (#2)

Auschwitz Lies: Legends, Lies and Prejudices on the Holocaust. By Carlo Mattogno and Germar Rudolf. The fallacious research and alleged "refutation" of Revisionist scholars by French biochemist G. Wellers (attacking Leuchter's famous report), Polish chemist Dr. J. Markiewicz and U.S. chemist Dr. Richard Green (taking on Rudolf's chemical research), Dr. John Zimmerman (tackling Mattogno on cremation issues), Michael Shermer and Alex Grobman (trying to prove it all), as well as researchers Keren, McCarthy and Mazal (who turned cracks into architectural features), are exposed for what they are: blatant and easily exposed political lies created to ostracize dissident historians. 4th ed., 420 pages, b&w illustrations, index. (#18)

Auschwitz: The Central Construction Office. By Carlo Mattogno. Ever since the Russian authorities granted western historians access to their state archives in the early 1990s, the files of the Central Construction Office of the Waffen-SS and Police Auschwitz, stored in a Moscow archive, have attracted the attention of scholars who are researching the history of this most infamous of all German war-time camps. Despite this interest, next to nothing has really been known so far about this very important office, which was responsible for the planning and construction of the Auschwitz camp complex, including the crematories which are said to have contained the "gas chambers." This emphasizes the importance of the present study, which not only sheds light into this hitherto hidden

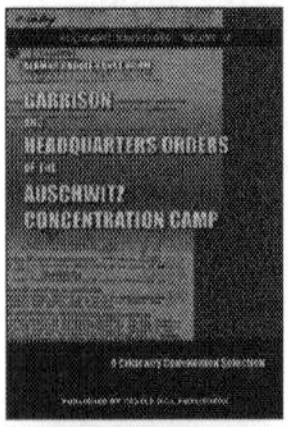

aspect of this camp's history, but also provides a deep understanding of the organization, tasks, and procedures of this office. 2nd ed., 188 pages, b&w illustrations, glossary, index. (#13)

Garrison and Headquarters Orders of the Auschwitz Camp. By Germar Rudolf and Ernst Böhm. A large number of all the orders ever issued by the various commanders of the infamous Auschwitz camp have been preserved. They reveal the true nature of the camp with all its daily events. There is not a trace in these orders pointing at anything sinister going on in this camp. Quite to the contrary, many orders are in clear and insurmountable contradiction to claims that prisoners were mass murdered, such as the children of SS men playing with inmates, SS men taking friends for a sight-seeing tour through the camp, or having a romantic stroll with their lovers around the camp grounds. This is a selection of the most pertinent of these orders together with comments putting them into their proper historical context. 185 pages, b&w ill., bibl., index (#34)

Special Treatment in Auschwitz: Origin and Meaning of a Term. By Carlo Mattogno. When appearing in German wartime documents, terms like "special treatment," "special action," and others have been interpreted as code words for mass murder. But that is not always true. This study focuses on documents about Auschwitz, showing that, while "special" had many different meanings, not a single one meant "execution." Hence the practice of deciphering an alleged "code language" by assigning homicidal meaning to harmless documents – a key component of mainstream historiography – is untenable. 2nd ed., 166 pages, b&w illustrations, bibliography, index. (#10)

Healthcare at Auschwitz. By Carlo Mattogno. In extension of the above study on *Special Treatment in Auschwitz*, this study proves the extent to which the German authorities at Auschwitz tried to provide health care for the inmates. Part 1 of this book analyzes the inmates' living conditions and the various sanitary and medical measures implemented. Part 2 explores what happened to registered inmates who were "selected" or subject to "special treatment" while disabled or sick. This study shows that a lot was tried to cure these inmates, especially under the aegis of Garrison Physician Dr. Wirths. Part 3 is dedicated to this very Dr. Wirths. His reality refutes the current stereotype of SS officers. 398 pages, b&w illustrations, bibliography, index. (#33)

Debunking the Bunkers of Auschwitz: Black Propaganda vs. History. By Carlo Mattogno. The "bunkers" at Auschwitz, two former farmhouses just outside the camp's perimeter, are claimed to have been the first homicidal gas chambers at Auschwitz specifically equipped for this purpose. With the help of original German wartime files as well as revealing air photos taken by Allied reconnaissance aircraft in 1944, this study shows that these homicidal "bunkers" never existed, how the rumors about them evolved as black propaganda created by resistance groups in the camp, and how this propaganda was transformed into a false reality. 2nd ed., 292 pages, b&w ill., bibliography, index. (#11)

Auschwitz: The First Gassing. Rumor and Reality. By Carlo Mattogno. The first gassing in Auschwitz is claimed to have occurred on Sept. 3, 1941 in a basement. The accounts reporting it are the archetypes for all later gassing accounts. This study analyzes all available sources about this alleged event. It shows that these sources contradict each other about the event's location, date, the kind of victims and their number, and many more aspects, which makes it impossible to extract a consistent story. Original wartime documents inflict a final blow to this legend and prove without a shadow of a doubt that this legendary event never happened. 3rd ed., 190 pages, b&w illustrations, bibliography, index. (#20)

Auschwitz: Crematorium I and the Alleged Homicidal Gassings. By Carlo Mattogno. The morgue of Crematorium I in Auschwitz is said to be the first homicidal gas chamber there. This study investigates all statements by witnesses and analyzes hundreds of wartime documents to accurately write a history of that building. Where witnesses speak of gassings, they are either very vague or, if specific, contradict one another and are refuted by documented and material facts. The author also exposes the fraudulent attempts of mainstream historians to convert the witnesses' black propaganda into "truth" by means of selective quotes, omissions, and distortions. Mattogno proves that this building's morgue was never a homicidal gas chamber, nor could it have worked as such. 2nd ed., 152 pages, b&w illustrations, bibliography, index. (#21)

Auschwitz: Open-Air Incinerations. By Carlo Mattogno. In spring and summer of 1944, 400,000 Hungarian Jews were deported to Auschwitz and allegedly murdered there in gas chambers. The Auschwitz crematoria are said to have been unable to cope with so many corpses. Therefore, every single day thousands of corpses are claimed to have been incinerated on huge pyres lit in deep trenches. The sky over Auschwitz was filled with thick smoke. This is what some witnesses want us to believe. This book examines the many testimonies regarding these incinerations and establishes whether these claims were even possible. Using air photos, physical evidence and wartime documents, the author shows that these claims are fiction. A new Appendix contains 3 papers on groundwater levels and cattle mass burnings. 2nd ed., 202 pages, b&w illustrations, bibliography, index. (#17)

The Cremation Furnaces of Auschwitz. By Carlo Mattogno & Franco Deana. An exhaustive study of the early history and technology of cremation in general and of the cremation furnaces of Auschwitz in particular. On a vast base of technical literature, extant wartime documents and material traces, the authors can establish the true nature and capacity of the Auschwitz cremation furnaces. They show that these devices were inferior makeshift versions of what was usually produced, and that their capacity to cremate corpses was lower than normal, too. This demonstrates that the Auschwitz crematoria were not evil facilities of mass destruction, but normal installations that barely managed to handle the victims among the inmates who died of various epidemics ravaging the camp throught its history. 2nd ed., 3 vols., 1201 pages, b&w and color illustrations (vols 2 & 3), bibliography, index, glossary. (#24)

Curated Lies: The Auschwitz Museum's Misrepresentations, Distortions and Deceptions. By Carlo Mattogno. Revisionist research results have put the Polish Auschwitz Museum under pressure to answer this challenge. In 2014, they answered with a book presenting documents allegedly proving their claims. But they cheated. In its main section, this study analyzes their "evidence" and reveals the appallingly mendacious attitude of the Auschwitz Museum authorities when presenting documents from their archives. This is preceded by a section focusing on the Auschwitz Museum's most-coveted asset: the alleged gas chamber inside the Old Crematorium, toured every year by well over a million visitors. *Curated Lies* exposes the many ways in which visitors have been deceived and misled by forgeries and misrepresentations about this building committed by the Auschwitz Museum, some of which are maintained to this day. 2nd ed., 259 pages, b&w illustrations, bibliography, index. (#38)

Deliveries of Coke, Wood and Zyklon B to Auschwitz: Neither Proof Nor Trace for the Holocaust. By Carlo Mattogno. Researchers from the Auschwitz Museum tried to prove the reality of mass extermination by pointing to documents about deliveries of wood and coke as well as Zyklon B to the Auschwitz Camp. If put into the actual historical and technical context, however, as is done by this study, these documents prove the exact opposite of what those orthodox researchers claim. 184 pages, b&w illust., bibl., index. (#40)

Mis-Chronicling Auschwitz. Danuta Czech's Flawed Methods, Lies and Deceptions in Her "Auschwitz Chronicle". By Carlo Mattogno. Danuta Czech's *Auschwitz Chronicle* is a reference book for the history of Auschwitz. Mattogno has compiled a long list of misrepresentations, outright lies and deceptions contained in it. This mega-fraud needs to be retired from the ranks of Auschwitz sources. Ca 320 pages, b&w illust., bibliography, index. (#47; Feb. 2022)

SECTION FOUR:
Witness Critique

Elie Wiesel, Saint of the Holocaust: A Critical Biography. By Warren B. Routledge. The world's first independent biography of Elie Wiesel shines the light of truth on this mythomaniac who has transformed the word "Holocaust" into the brand name of the world's greatest hoax. Here, both Wiesel's personal deceits and the whole myth of "the six million" are laid bare for the reader's perusal. It shows how Zionist control of the U.S. Government as well as the nation's media and academic apparatus has allowed Wiesel and his fellow extremists to force a string of U.S. presidents to genuflect before this imposter as symbolic acts of subordination to World Jewry, while simultaneously forcing school children to submit to Holocaust brainwashing by their teachers. 3rd ed., 458 pages, b&w illustration, bibliography, index. (#30)

Auschwitz: Eyewitness Reports and Perpetrator Confessions. By Jürgen Graf. The traditional narrative

of what transpired at the infamous Auschwitz Camp during WWII rests almost exclusively on witness testimony. This study critically scrutinizes the 30 most-important of them by checking them for internal coherence, and by comparing them with one another as well as with other evidence such as wartime documents, air photos, forensic research results, and material traces. The result is devastating for the traditional narrative. 372 pages, b&w illust., bibl., index. (#36)

Commandant of Auschwitz: Rudolf Höss, His Torture and His Forced Confessions. By Carlo Mattogno & Rudolf Höss. From 1940 to 1943, Rudolf Höss was the commandant of the infamous Auschwitz Camp. After the war, he was captured by the British. In the following 13 months until his execution, he made 85 depositions of various kinds in which he confessed his involvement in the "Holocaust." This study first reveals how the British tortured him to extract various "confessions." Next, all of Höss's depositions are analyzed by checking his claims for internal consistency and comparing them with established historical facts. The results are eye-opening... 2nd ed., 411 pages, b&w illust., bibliography, index. (#35)

An Auschwitz Doctor's Eyewitness Account: The Tall Tales of Dr. Mengele's Assistant Analyzed. By Miklos Nyiszli & Carlo Mattogno. Nyiszli, a Hungarian physician, ended up at Auschwitz in 1944 as Dr. Mengele's assistant. After the war he wrote a book and several other writings describing what he claimed to have experienced. To this day some traditional historians take his accounts seriously, while others reject them as grotesque lies and exaggerations. This study presents and analyzes Nyiszli's writings and skillfully separates truth from fabulous fabrication. 2nd ed., 484 pages, b&w illustrations, bibliography, index. (#37)

Rudolf Reder versus Kurt Gerstein: Two False Testimonies on the Bełżec Camp Analyzed. By Carlo Mattogno. Only two witnesses have ever testified substantially about the alleged Belzec Extermination Camp: The survivor Rudolf Reder and the SS officer Kurt Gerstein. Gerstein's testimonies have been a hotspot of revisionist critique for decades. It is now discredited even among orthodox historians. They use Reder's testimony to fill the void, yet his testimonies are just as absurd. This study thoroughly scrutinizes Reder's various statements, critically revisits Gerstein's various depositions, and then compares these two testimonies which are at once similar in some respects, but incompatible in others. 216 pages, b&w illust., bibliography, index. (#43)

Sonderkommando Auschwitz I: Nine Eyewitness Testimonies Analyzed. By Carlo Mattogno. To this day, the 1979 book *Auschwitz Inferno* by alleged former Auschwitz "Sonderkommando" member Filip Müller has a great influence on the perception of Auschwitz by the public and by historians. This book critically analyzes Müller's various post-war statements, which are full of exaggerations, falsehoods and plagiarized text passages. Also scrutinized are the testimonies of eight other claimed former *Sonderkommando* members: D. Paisikovic, S. Jankowski, H. Mandelbaum, L. Nagraba, J. Rosenblum, A. Pilo, D. Fliamenbaum and S. Karolinskij. 304 pages, b&w illust., bibliography, index. (#44)

Sonderkommando Auschwitz II: The False Testimonies by Henryk Tauber and Szlama Dragon. By Carlo Mattogno. Auschwitz survivor and former member of the so-called "Sonderkommando" Henryk Tauber is one of the most important witnesses about the alleged gas chambers inside the crematoria at Auschwitz, because right at the war's end, he made several extremely detailed depositions about it. The same is true for Szlama Dragon, only he claims to have worked at the so-called "bunkers" of Birkenau, two makeshift gas chambers just outside the camp perimeter. This study thoroughly scrutinizes these two key testimonies. Ca. 250 pages, b&w illust., bibliography, index. (#45; May 2022)

Sonderkommando Auschwitz III: We Wept Crocodile Tears. By Carlo Mattogno. This book focuses on the critical analysis of witness testimonies on the alleged Auschwitz gas chambers recorded or published in the 1990s and early 2000s, such as J. Sackar, A. Dragon, J. Gabai, S. Chasan, L. Cohen and S. Venezia, among others. Ca 250 pages, b&w illust., bibliography, index. (#46, late 2022)

For current prices and availability, and to learn more, go to www.HolocaustHandbooks.com – by simply scanning the QR code to the left.
Published by Castle Hill Publishers, PO Box 243, Uckfield, TN22 9AW, UK

Books by and from Castle Hill Publishers

Below please find some of the books published or distributed by Castle Hill Publishers in the United Kingdom. For our current and complete range of products visit our web store at shop.codoh.com.

The Holocaust: An Introduction. By Thomas Dalton. The Holocaust was perhaps the greatest crime of the 20th Century. Six million Jews, we are told, died by gassing, shooting, and deprivation. But: Where did the six-million figure come from? How, exactly, did the gas chambers work? Why do we have so little physical evidence from major death camps? Why haven't we found even a fraction of the six million bodies, or their ashes? Why has there been so much media suppression and governmental censorship on this topic? In a sense, the Holocaust is the greatest murder mystery in history. It is a topic of greatest importance for the present day. Let's explore the evidence, and see where it leads. 128 pp. pb, 5"×8", ill., bibl., index.

Auschwitz: A Three-Quarter Century of Propaganda: Origins, Development and Decline of the "Gas Chamber" Propaganda Lie. By Carlo Mattogno. During the war, wild rumors were circulating about Auschwitz: Germans testing war gases; inmates murdered in electrocution chambers, with gas showers or pneumatic hammers; living people sent on conveyor belts into furnaces; oils, grease and soap made of the victims. Nothing of it was true. When the Soviets captured Auschwitz in early 1945, they reported that 4 million inmates were killed on electrocution conveyor belts discharging their load directly into furnaces. That wasn't true either. After the war, "witnesses" and "experts" repeated these claims and added more: mass murder with gas bombs, gas chambers made of canvas; carts driving living people into furnaces; crematoria burning 400 million victims... Again, none of it was true. This book gives an overview of the many rumors, myths and lies about Auschwitz today rejected as untrue. It then explains by which ridiculous methods some claims were accepted and turned into "history," although they are just as untrue. 125 pp. pb, 5"×8", ill., bibl., index, b&w ill.

Auschwitz: A Judge Looks at the Evidence. By Wilhelm Stäglich. Auschwitz is the epicenter of the Holocaust, where more people are said to have been murdered than anywhere else. This claim is based on a wide range of evidence, the most important of which was presented during two trials: the International Military Tribunal of 1945/46, and the German Auschwitz Trial of 1963-1965 in Frankfurt. In this book, Wilhelm Stäglich, a former German judge, critically analyzes this evidence. He reveals the incredibly scandalous way in which Allied victors and German courts bent and broke the law in order to come to politically foregone conclusions. Stäglich also exposes the superficial way in which historians are dealing with the many incongruities and discrepancies of the historical record. 3rd edition 2015, 422 pp. pb, 6"×9", b&w ill.

Tell the Truth & Shame the Devil. By Gerard Menuhin. This Jewish author says the "Holocaust" is a wartime-propaganda myth turned into an extortion racket. Far from bearing the sole guilt for starting WWII as alleged at Nuremberg, Germany is mostly innocent and made numerous attempts to avoid and later to end the confrontation. During the 1930s, Germany was confronted by a powerful Jewish-dominated world plutocracy out to destroy it... Yes, a Jew says all this. The author is the son of the great US-born violinist Yehudi Menuhin, who, though from a long line of rabbinical ancestors, fiercely criticized the foreign policy of Israel and its repression of the Palestinians. 4th edition 2017, 432 pp. pb, 6"×9", b&w ill.

Exactitude: Festschrift for Prof. Dr. Robert Faurisson. By R.H. Countess, C. Lindtner, G. Rudolf (eds.) Faurisson probably deserves the title of the most-courageous intellectual of the 20th and the early 21st Century. With bravery and steadfastness, he challenged the dark forces of historical and political fraud with his unrelenting exposure of their lies and hoaxes surrounding the orthodox Holocaust narrative. This book describes and celebrates the man and his work dedicated to accuracy and marked by insubmission. 146 pp. pb, 6"×9", b&w ill.

Auschwitz – Forensically Examined. By Cyrus Cox. Modern forensic crime-scene investigations can reveal a lot about the Holocaust. There are many big tomes about this. But if you want it all in a nutshell, read this booklet. It condenses the most-important findings of Auschwitz forensics into a quick and easy read. In the first section, the forensic investigations conducted so far are reviewed. In the second section, the most-important results of these studies are summarized. The main arguments focus on two topics. The first centers around the poison allegedly used at Auschwitz for mass murder: Zyklon B. Did it leave any traces in masonry where it was used? Can it be detected to this day? The second topic deals with mass cremations. Did the crematoria of Auschwitz have the claimed huge capacity? Do air photos taken during the war confirm witness statements on huge smoking pyres? This book gives the answers, together with many references to source material and further reading. The third section reports on how the establishment has reacted to these research results. 124 pp. pb., 5“×8“, b&w ill., bibl., index

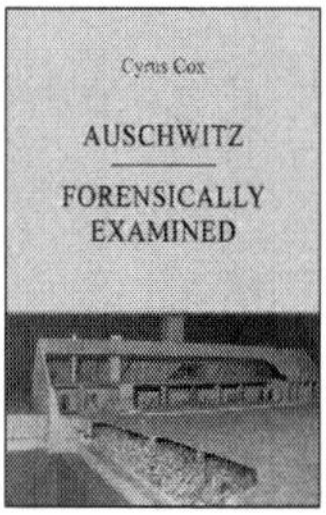

The Second Babylonian Captivity: The Fate of the Jews in Eastern Europe since 1941. By Steffen Werner. “But if they were not murdered, where did the six million deported Jews end up?” This is a standard objection to the revisionist thesis that the Jews were not killed in extermination camps. It demands a well-founded response. While researching an entirely different topic, Steffen Werner accidentally stumbled upon the most-peculiar demographic data of Byelorussia. Years of research subsequently revealed more and more evidence which eventually allowed him to substantiate a breathtaking and sensational proposition: The Third Reich did indeed deport many of the Jews of Europe to Eastern Europe in order to settle them there “in the swamp.” This book, first published in German in 1990, was the first well-founded work showing what really happened to the Jews deported to the East by the National Socialists, how they have fared since, and who, what and where they are “now” (1990). It provides context and purpose for hitherto-obscure and seemingly random historical events and quite obviates all need for paranormal events such as genocide, gas chambers, and all their attendant horrifics. With a preface by Germar Rudolf with references to more-recent research results in this field of study confirming Werner's thesis. 190 pp. pb, 6”×9”, b&w ill., bibl., index

Holocaust Skepticism: 20 Questions and Answers about Holocaust Revisionism. By Germar Rudolf. This 15-page brochure introduces the novice to the concept of Holocaust revisionism, and answers 20 tough questions, among them: What does Holocaust revisionism claim? Why should I take Holocaust revisionism more seriously than the claim that the earth is flat? How about the testimonies by survivors and confessions by perpetrators? What about the pictures of corpse piles in the camps? Why does it matter how many Jews were killed by the Nazis, since even 1,000 would have been too many? ... Glossy full-color brochure. PDF file free of charge available at www.HolocaustHandbooks.com, Option “Promotion”. This item is *not* copyright-protected. Hence, you can do with it whatever you want: download, post, email, print, multiply, hand out, sell... 15 pp., stapled, 8.5“×11“, full-color throughout.

Bungled: “Denying the Holocaust” How Deborah Lipstadt Botched Her Attempt to Demonstrate the Growing Assault on Truth and Memory. By Germar Rudolf. With her book *Denying the Holocaust*, Deborah Lipstadt tried to show the flawed methods and extremist motives of “Holocaust deniers.” This book demonstrates that Dr. Lipstadt clearly has neither understood the principles of science and scholarship, nor has she any clue about the historical topics she is writing about. She misquotes, mistranslates, misrepresents, misinterprets, and makes a plethora of wild claims without backing them up with anything. Rather than dealing thoroughly with factual arguments, Lipstadt's book is full of *ad hominem* attacks on her opponents. It is an exercise in anti-intellectual pseudo-scientific arguments, an exhibition of ideological radicalism that rejects anything which contradicts its preset conclusions. **F for FAIL.** 2nd ed., 224 pp. pb, 5“×8“, bibl., index, b&w ill.

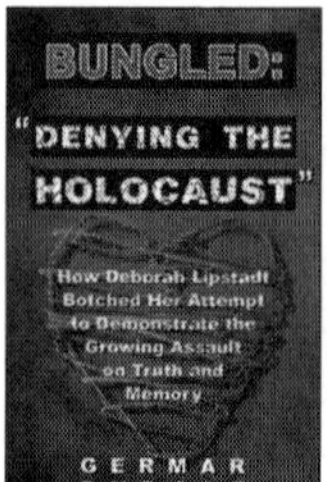

Bungled: "Denying History". How Michael Shermer and Alex Grobman Botched Their Attempt to Refute Those Who Say the Holocaust Never Happened. By Carolus Magnus (Carlo Mattogno). *Skeptic Magazine* editor Michael Shermer and Alex Grobman from the Simon Wiesenthal Center wrote a book in 2000 which they claim is "a thorough and thoughtful answer to all the claims of the Holocaust deniers." In 2009, a new "updated" edition appeared with the same ambitious goal. In the meantime, revisionists had published some 10,000 pages of archival and forensic research results. Would their updated edition indeed answer all the revisionist claims? In fact, Shermer and Grobman completely ignored the vast amount of recent scholarly studies and piled up a heap of falsifications, contortions, omissions, and fallacious interpretations of the evidence. Finally, what the authors claim to have demolished is not revisionism but a ridiculous parody of it. They ignored the known unreliability of their cherry-picked selection of evidence, utilizing unverified and incestuous sources, and obscuring the massive body of research and all the evidence that dooms their project to failure. **F for FAIL.** 162 pp. pb, 5"×8", bibl., index, b&w ill.

Bungled: "Debunking Holocaust Denial Theories". How James and Lance Morcan Botched Their Attempt to Affirm the Historicity of the Nazi Genocide. By Carolus Magnus (Carlo Mattogno). The novelists and moviemakers James and Lance Morcan have produced a book "to end [Holocaust] denial once and for all." To do this, "no stone was left unturned" to verify historical assertions by presenting "a wide array of sources" meant "to shut down the debate deniers wish to create. One by one, the various arguments Holocaust deniers use to try to discredit wartime records are carefully scrutinized and then systematically disproven." It's a lie. First, the Morcans completely ignored the vast amount of recent scholarly studies published by revisionists; they didn't even mention them. Instead, they engaged in shadowboxing, creating some imaginary, bogus "revisionist" scarecrow which they then tore to pieces. In addition, their knowledge even of their own side's source material was dismal, and the way they backed up their misleading or false claims was pitifully inadequate. **F for FAIL.** 144 pp. pb, 5"×8", bibl., index, b&w ill.

Stalin's War of Extermination 1941-1945. By Joachim Hoffmann. A German government historian documents Stalin's murderous war against the German army and the German people. Based on the author's lifelong study of German and Russian military records, this book reveals the Red Army's grisly record of atrocities against soldiers and civilians, as ordered by Stalin. Since the 1920s, Stalin planned to invade Western Europe to initiate the "World Revolution." He prepared an attack which was unparalleled in history. The Germans noticed Stalin's aggressive intentions, but they underestimated the strength of the Red Army. What unfolded was the cruelest war in history. This book shows how Stalin and his Bolshevik henchman used unimaginable violence and atrocities to break any resistance in the Red Army and to force their unwilling soldiers to fight against the Germans. The book explains how Soviet propagandists incited their soldiers to unlimited hatred against everything German, and he gives the reader a short but extremely unpleasant glimpse into what happened when these Soviet soldiers finally reached German soil in 1945: A gigantic wave of looting, arson, rape, torture, and mass murder... **428 pp. pb, 6"×9", bibl., index, b&w ill.**

Who Started World War II: Truth for a War-Torn World. By Udo Walendy. For seven decades, mainstream historians have insisted that Germany was the main, if not the sole culprit for unleashing World War II in Europe. In the present book this myth is refuted. There is available to the public today a great number of documents on the foreign policies of the Great Powers before September 1939 as well as a wealth of literature in the form of memoirs of the persons directly involved in the decisions that led to the outbreak of World War II. Together, they made possible Walendy's present mosaic-like reconstruction of the events before the outbreak of the war in 1939. This book has been published only after an intensive study of sources, taking the greatest care to minimize speculation and inference. The present edition has been translated completely anew from the German original and has been slightly revised. 500 pp. pb, 6"×9", index, bibl., b&w ill.

Resistance Is Obligatory! By Germar Rudolf. In 2005 Rudolf, a peaceful dissident and publisher of revisionist literature, was kidnapped by the U.S. government and deported to Germany. There the local lackey regime staged a show trial against him for his historical writings. Rudolf was not permitted to defend his historical opinions, as the German penal law prohibits this. Yet he defended himself anyway: For 7 full days Rudolf gave a speech in the courtroom, during which he proved systematically that only the revisionists are scholarly in their approach, whereas the Holocaust orthodoxy is merely pseudo-scientific. He then explained in detail why it is everyone's obligation to resist, without violence, a government which throws peaceful dissidents into dungeons. When Rudolf tried to publish his public defence speech as a book from his prison cell, the public prosecutor initiated a new criminal investigation against him. After his probation time ended in 2011, he dared publish this speech anyway... 2nd ed. 2016, 378 pp. pb, 6"×9", b&w ill.

Hunting Germar Rudolf: Essays on a Modern-Day Witch Hunt. By Germar Rudolf. German-born revisionist activist, author and publisher Germar Rudolf describes which events made him convert from a Holocaust believer to a Holocaust skeptic, quickly rising to a leading personality within the revisionist movement. This in turn unleashed a tsunami of persecution against him: lost his job, denied his PhD exam, destruction of his family, driven into exile, slandered by the mass media, literally hunted, caught, put on a show trial where filing motions to introduce evidence is illegal under the threat of further prosecution, and finally locked up in prison for years for nothing else than his peaceful yet controversial scholarly writings. In several essays, Rudolf takes the reader on a journey through an absurd world of government and societal persecution which most of us could never even fathom actually exists in a "Western democracy"... 304 pp. pb, 6"×9", bibl., index, b&w ill.

The Day Amazon Murdered History. By Germar Rudolf. Amazon is the world's biggest book retailer. They dominate the U.S. and several foreign markets. Pursuant to the 1998 declaration of Amazon's founder Jeff Bezos to offer "the good, the bad and the ugly," customers once could buy every title that was in print and was legal to sell. However, in early 2017, a series of anonymous bomb threats against Jewish community centers occurred in the U.S., fueling a campaign by Jewish groups to coax Amazon into banning revisionist writings, falsely portraying them as anti-Semitic. On March 6, 2017, Amazon caved in and banned more than 100 books with dissenting viewpoints on the Holocaust. In April 2017, an Israeli Jew was arrested for having placed the fake bomb threats, a paid "service" he had offered for years. But that did not change Amazon's policy. Its stores remain closed for history books Jewish lobby groups disapprove of. This book accompanies the documentary of the same title. Both reveal how revisionist publications had become so powerfully convincing that the powers that be resorted to what looks like a dirty false-flag operation in order to get these books banned from Amazon... 128 pp. pb, 5"×8", bibl., b&w ill.

Hitler's Revolution: Ideology, Social Programs, Foreign Affairs. By Richard Tedor. Defying all boycotts, Adolf Hitler transformed Germany from a bankrupt state to the powerhouse of Europe within just four years, thus becoming Germany's most popular leader ever. How was this possible? This study tears apart the dense web of calumny surrounding this controversial figure. It draws on nearly 200 published German sources, many from the Nazi era, as well as documents from British, U.S., and Soviet archives that describe not only what Hitler did but, more importantly, why he did it. These sourcs also reveal the true war objectives of the democracies – a taboo subject for orthodox historians – and the resulting world war against Germany. This book is aimed at anyone who feels that something is missing from conventional accounts. 2nd ed., 309 pp. pb, 6"×9", index, bibl.

Hitler on the Jews. By Thomas Dalton. That Adolf Hitler spoke out against the Jews is beyond obvious. But of the thousands of books and articles written on Hitler, virtually none quotes Hitler's exact words on the Jews. The reason for this is clear: Those in positions of influence have incentives to present a simplistic picture of Hitler as a blood-thirsty tyrant. However, Hitler's take on the Jews is far more complex and sophisticated. In this book, for the first time, you can make up your own mind by reading nearly every idea that Hitler put forth about the Jews, in considerable detail and in full context. This is the first book ever to compile his remarks on the Jews. As you will discover, Hitler's analysis of the Jews, though hostile, is erudite, detailed, and – surprise, surprise – largely aligns with events of recent decades. There are many lessons here for the modern-day world to learn. 200 pp. pb, 6"×9", index, bibl.

Goebbels on the Jews. By Thomas Dalton. From the age of 26 until his death in 1945, Joseph Goebbels kept a near-daily diary. From it, we get a detailed look at the attitudes of one of the highest-ranking men in Nazi Germany. Goebbels shared Hitler's dislike of the Jews, and likewise wanted them totally removed from the Reich territory. Ultimately, Goebbels and others sought to remove the Jews completely from the Eurasian land mass—perhaps to the island of Madagascar. This would be the "final solution" to the Jewish Question. Nowhere in the diary does Goebbels discuss any Hitler order to kill the Jews, nor is there any reference to extermination camps, gas chambers, or any methods of systematic mass-murder. Goebbels acknowledges that Jews did indeed die by the thousands; but the range and scope of killings evidently fall far short of the claimed figure of 6 million. This book contains, for the first time, every significant diary entry relating to the Jews or Jewish policy. Also included are partial or full transcripts of 10 major essays by Goebbels on the Jews. 274 pp. pb, 6"×9", index, bibl.

The Jewish Hand in the World Wars. By Thomas Dalton. For many centuries, Jews have had a negative reputation in many countries. The reasons given are plentiful, but less-well-known is their involvement in war. When we examine the causal factors for wars, and look at their primary beneficiaries, we repeatedly find a Jewish presence. Throughout history, Jews have played an exceptionally active role in promoting and inciting wars. With their long-notorious influence in government, we find recurrent instances of Jews promoting hard-line stances, being uncompromising, and actively inciting people to hatred. Jewish misanthropy, rooted in Old Testament mandates, and combined with a ruthless materialism, has led them, time and again, to instigate warfare if it served their larger interests. This fact explains much about the present-day world. In this book, Thomas Dalton examines in detail the Jewish hand in the two world wars. Along the way, he dissects Jewish motives and Jewish strategies for maximizing gain amidst warfare, reaching back centuries. 2nd ed., 231 pp. pb, 6"×9", index, bibl.

Eternal Strangers: Critical Views of Jews and Judaism through the Ages. By Thomas Dalton. It is common knowledge that Jews have been disliked for centuries. But why? Our best hope for understanding this recurrent 'anti-Semitism' is to study the history: to look at the actual words written by prominent critics of the Jews, in context, and with an eye to any common patterns that might emerge. Such a study reveals strikingly consistent observations: Jews are seen in very negative, yet always similar terms. The persistence of such comments is remarkable and strongly suggests that the cause for such animosity resides in the Jews themselves—in their attitudes, their values, their ethnic traits and their beliefs.. This book addresses the modern-day "Jewish problem" in all its depth—something which is arguably at the root of many of the world's social, political and economic problems. 186 pp. pb, 6"×9", index, bibl.

Streicher, Rosenberg, and the Jews: The Nuremberg Transcripts. By Thomas Dalton. Who, apart from Hitler, contrived the Nazi view on the Jews? And what were these master ideologues thinking? During the post-war International Military Tribunal at Nuremberg, the most-interesting men on trial regarding this question were two with a special connection to the "Jewish Question": Alfred Rosenberg and Julius Streicher. The cases against them, and their personal testimonies, examined for the first time nearly all major aspects of the Holocaust story: the "extermination" thesis, the gas chambers, the gas vans, the shootings in the East, and the "6 million." The truth of the Holocaust has been badly distorted for decades by the powers that be. Here we have the rare opportunity to hear firsthand from two prominent figures in Nazi Germany. Their voices, and their verbatim transcripts from the IMT, lend some much-needed clarity to the situation. 330 pp. pb, 6"×9", index, bibl.

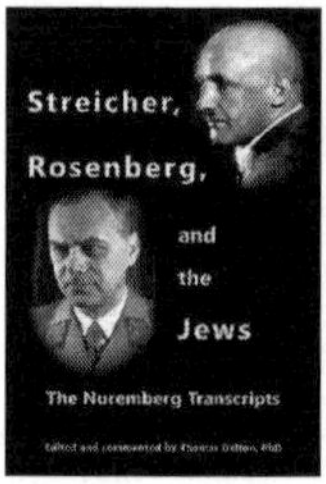

The First Zündel Trial: The Transcript. In the early 1980s, Ernst Zündel, a German immigrant living in Toronto, was indicted for allegedly spreading "false news" by selling copies of Richard Harwood's brochure *Did Six Million Really Die?,* which challenged the accuracy of the orthodox Holocaust narrative. When the case went to court in 1985, so-called Holocaust experts and "eyewitnesses" of the alleged homicidal gas chambers at Auschwitz were cross-examined for the first time in history by a competent and skeptical legal team. The results were absolutely devastating for the Holocaust orthodoxy. Even the prosecutor, who had summoned these witnesses to bolster the mainstream Holocaust narrative, became at times annoyed by their incompetence and mendacity. For decades, these mind-boggling trial transcripts were hidden from public view. Now, for the first time, they have been published in print in this new book – unabridged and unedited. 820 pp. pb, 8.5"×11"

The Second Zündel Trial: Excerpts from the Transcript. By Barbara Kulaszka (ed.). In 1988. German-Canadian Ernst Zündel was on trial for a second time for allegedly spreading "false news" about the Holocaust. Zündel staged a magnificent defense in an attempt to prove that revisionist concepts of "the Holocaust" are essentially correct. Although many of the key players have since passed away, including Zündel, this historic trial keeps having an impact. It inspired major research efforts as expounded in the series *Holocaust Handbooks*. In contrast to the First Zündel Trial of 1985, the second trial had a much greater impact internationally, mainly due to the ***Leuchter Report***, the first independent forensic research performed on Auschwitz, which was endorsed on the witness stand by British bestselling historian David Irving. The present book features the essential contents of this landmark trial with all the gripping, at-times-dramatic details. When Amazon.com decided to ban this 1992 book on a landmark trial about the "Holocaust", we decided to put it back in print, lest censorship prevail... 498 pp. pb, 8.5"×11", bibl., index, b&w ill.

Lies & Gravy: Landmarks in Human Decay – Two Plays. By Gerard Menuhin. A long time ago, in a galaxy far, far away, the hallucination of global supremacy was born. Few paid it any attention. After centuries of interference, when the end is in sight, we're more inclined to take it seriously. But now, we have only a few years of comparative freedom left before serfdom submerges us all. So it's time to summarize our fall and to name the guilty, or, as some have it, to spot the loony. Sometimes the message is so dire that the only way to get it across is with humor – to act out our predicament and its causes. No amount of expert testimony can match the power of spectacle. Here are a few of the most-telling stages in the chosenites' crusade against humanity, and their consequences, as imagined by the author. We wonder whether these two consecutive plays will ever be performed onstage... 112 pp. pb, 5"×8"

For current prices and availability see book-finder sites such as www.findbookprices.com; learn more at https://shop.codoh.com by simply scanning the QR code to the left with you smart device.
Published by Castle Hill Publishers, PO Box 243, Uckfield, TN22 9AW, UK